Cultures@SiliconValley

Cultures@SiliconValley

J. A. English-Lueck

Stanford University Press
Stanford, California
2002

Stanford University Press
Stanford, California
© 2002 by the Board of Trustees of the
Leland Stanford Junior University
Printed in the United States of America

ISBN 0-8047-4428-9 (alk. paper)
ISBN 0-8047-4429-7 (pbk. : alk. paper)

This book is printed on acid-free, archival-quality paper.

Original printing 2002
Last figure below indicates year of this printing:
11   10   09   08   07   06   05   04   03   02

Typeset in 10/13 Sabon

To Karl, who chose the storm over the calm,
and our spirited daughters—Miriam and Eilene.

Acknowledgments

This book draws on nearly a decade of ethnographic work in Silicon Valley. The book you are reading comes from data gathered by researchers in interviews, observations, or, occasionally, their own life experience. Hundreds of people have granted us access to their words and their experiences. Most of those people did so with the understanding that their anonymity would be preserved. The names of individuals, and often organizations, have been changed to preserve that confidentiality. We owe a debt of gratitude to those people who submitted to our interviews and observations for their patience and generosity.

The ethnographic researchers include the author, Charles Darrah, James Freeman, and a small horde of anthropologists-in-training. Many of the best ideas in this book came from project meetings and corridor talk with Chuck and Jim. I owe them my deepest thanks. I also want to thank students from the following anthropology courses—Ethnographic Methods, Culture and Conflict, Culture and Personality, Wealth and Power, and Emerging Global Cultures. Students in these courses helped collect data on the community and shared events from their own lives.

The compilation of the ethnographic database on Silicon Valley would not be possible without the efforts of Teresa Bennett, Lauralee Brown, Lorraine Burgman, Mary Cashion, Rachel Caso, T. C. Chang, David Cismowski, Sharon Covarrubias, Blair M. Dunton, Joseph Duran, Jennie Eaton, Debbie Faires, Elan Finch, Esther Foley, Vicki Geissinger, Joe Hertzbach, Jennifer Holforty, Veronica Keiffer, Robyn Lauziere, Kathleen MacKenzie, Piper McNulty, Eric Metz, Robert Olds, Doris O'Loughlin, Dana Ou, Diana M. Petry, Naftoli Pickard, Norma Rivera, Linda Quach, Eric Rhebergen, Paula Rockstroh, Jason Scatena, Jason Silz, Liz Snyder, Joelle Sorensen, Lydia Struich, Maho Teraguchi, Janet Thieman, Rhonda Vague, Araceli Valle, Robin Velte, Susan Weatherly, and Israel Zuckerman. Special thanks go to our heroic staff: Bonnie Evans, who coded massive amounts of data, and Deborah Dalton, project archivist. Karl Lueck, our data wrangler, graphics artist, and archivist-at-large, deserves praise both for his professional competence and his forbearance as a supportive spouse. I also want to thank Thomas Boures for his generosity in sharing his photographic portfolio.

I thank my children—Miriam and Eilene Lueck—for their patience when mother was off doing fieldwork or chained to the computer. I also want to thank the families of Chuck and Jim—Janice, Zachary and Joshua Darrah, Pat and Karsten Freeman—for their sacrifices during the years of data collection. I also want to thank the English and Lueck families, and my friends Eilene Cross, C. Joan Sutherland, Thomas Layton, Margaret Graham, and Russell Skowronek, whose optimistic spirits kept me going while I simultaneously struggled to teach, research, and administer. I also offer a special thanks to all the ladies of "The List," especially Barbara and Elizabeth, my online friends who so generously gave me their support and encouragement.

I must also acknowledge my institutional supporters, including San Jose State University's College of Social Science, the National Science Foundation, and the Alfred P. Sloan Foundation. I especially want to thank Sloan Foundation's Kathleen Christensen, who supported our research on the interaction of work and family among middle-class Americans. The Institute for the Future has been our primary collaborator—especially Andrea Saveri, Rod Falcon, Paul Saffo, Marina Gorbis, and Bob Johanson. I am grateful for the invaluable assistance of Vinay Kumar, director of the Centre for American Education, Peter Liu of W. I. Harper, and Daragh Kelly of Forfas for their help in our fieldwork in India, Taiwan, and Ireland. The research team at San Jose State has also collaborated with several other partners, coordinating our research goals with theirs. The American Anthropological Association (especially Dr. Peggy Overbey), Daimler-Benz, Ericsson, Pitney-Bowes, MIT/Working Partnerships, Xerox PARC, the Intel Architecture Labs, Interval Research, Sapient, the Tech Museum of Innovation, and Joint Ventures:Silicon Valley Network have all provided inspiration by exchanging research ideas. I also want to thank the many journalists who took the time to ask pointed questions and listen as I navigated my way through mountains of information. And I wish to recognize the tireless efforts of Ruhama Veltfort, my editorial consultant, and Nathan MacBrien, Muriel Bell, John Feneron, Martin Hanft, and Sumathi Raghavan from Stanford University Press.

Contents

4 pages of photographs follow page 44

Cultures@SiliconValley

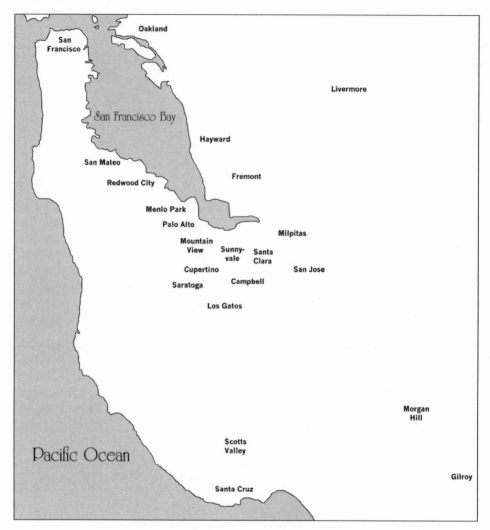

The Silicon Valley Region

Prologue

The Morning Report

Asok is a software engineer in one of the showcase companies in Silicon Valley, a gleaming edifice of glass and tile. He and his wife were born in India. Like many in his network of friends, he went to Stanford University to pursue a graduate degree and found work in a large company. He stayed in that company for three years, coding, learning American ways, and discovering that political hierarchies in American workplaces are very different from those he had known in India. Eager and enthusiastic, he put his heart and soul into his first project. His teammates were like family. They worked long hours together, eighty to a hundred hours per week in the crunch times. They went to Burgess Park for picnics, and loaned each other money. It was a heady experience. Suddenly, the product on which his project was based was canceled. His group was broken up and his teammates were distributed among a number of other projects. It was a character-building experience and Asok struggled with his grief.

His next project was less glamorous, and ultimately less satisfying: writing code for an upgrade to a widely available graphics program. Meanwhile, he married, bought a condo, and kept up his ties with classmates from India and Stanford, and with his old project team. Almost as in an apprenticeship, he stayed with the company until he had "upgraded his skill set." Then, leaping into the furnace, he joined an old classmate in a start-up. They were determined to make it but found it difficult to develop their many ideas into a product a client would be willing to buy. After two years of not quite making it to the Initial Public Offering stage that would make them wealthy, Asok began to rethink whether he should continue to work more than a hundred hours a week at a failing enterprise. He then heard about another position, via a buddy from his old institute in Haryana, India, who also worked in Silicon Valley. He took the job, and found both the work and the organization congenial and challenging. At the moment, his situation is relatively stable, but he knows that at any time a management decision in the now vast organization could jeopardize his status. Oh well, there are many more jobs out there.

Priyesh and his wife, Sima, old friends of Asok's, are having breakfast. Sima tells Priyesh that her relatives are asking if they will go back to India soon. They shake their heads, wondering why they would want to go back. She says, "I do not miss India . . . anything Indian I want I have it right here—grocery stores, temples, cultural programs, Hindu magazines. There are three movie theaters in the Bay Area showing Hindi movies seven days a week. I only miss my family." Priyesh reflects that here in Silicon Valley he can have all the good parts of India—the people and culture—without having to put up with a decaying bureaucracy and failing infrastructure. They argue gently about whether they really live in an Indian world. She points out that while many of their friends are Indian, including a recently immigrated cousin, Priyesh still has to work with many non-Indians.

Sima adds that she has to interact with very different cultures in their son's preschool—where they celebrate Chinese New Year and Cinco de Mayo. She pauses and then notes that it isn't really too different from interacting with all the different religions and cultures back in India. Priyesh scoffs at the whole problem. All these cultural differences don't matter to him. When he is at work, it is the technology that matters. Whether you are Irish, Chinese, or from New York just isn't important. Well, maybe a bit. The Irish speak better English and he can understand them more easily. What he really doesn't understand are the locally born folks who act as if Silicon Valley was their own invention. Maybe in the beginning this was true, but now it just isn't so simple. Sima responds enthusiastically, noting, "All the different engineers from different countries . . . that is why the brain power is there. If it was just Indians I don't think it would have happened. If it was just Taiwanese, just Chinese, it would not have happened. It's all the different ethnic groups that come together. All these engineers come from different countries and that's why it's making Silicon Valley successful today!"

Priyesh goes back into his home office. He makes a practice of telecommuting every day from about six to nine o'clock in the morning, to avoid the worst freeway traffic. His computer at home is nearly an exact duplicate of his computer at work. The computer is his, but the company provides the infrastructure that allows him to connect to the network. He can, at the very least, spend those hours reviewing his e-mail. He gets from fifty to one hundred e-mails each day. Many are work-related, but others connect him to mailing lists of people who are interested in technology stock investment, or people who like to play Indian music or tennis. They will send each other quick messages to set up meeting times or to announce the arrival of a particular artist. He also stays in touch with his family in India with e-mail, since that is the most convenient way to communicate across time zones. In addition to communication, Priyesh uses the morning "chunk" to get the information he needs in order to "set up" his software tests remotely. The network allows him to "work" in the computer laboratory from the convenience of his home or cubicle. When he is in his cubicle, he does use his phone from time to time to talk to his wife, but within his company, life is lived on the computer.

Across the street from the home where Priyesh telecommutes from, Heidi, one of those "locally born people," begins her day. She is almost twenty and Silicon Valley is her heritage. Her father has worked for a large technical company for most of her conscious life. She loves God and photography, and works with her friends on a magazine covering alternative music and extreme sports. It is a great job for her while she is a student at De Anza Community College. She learned her sophisticated computer skills at her father's knee. She watched him work his network, a broad spectrum of friends, to find whatever bit of technical information he needed. In addition to producing the graphics for the magazine, she markets it, never really needing to go beyond her network of church and family friends, school mates, fellow musicians and kin. She doesn't think there is anything revolutionary about Silicon Valley. Naturally, she will use technical skills to develop her own business. If this gig doesn't work, something else will come up. It always does. She wishes it were cheaper to live here—she would like to have her own apartment—but that is just impossible. But living with her parents is okay. Dad gave her his old computer when he upgraded and when she needs more RAM, he lets her use his new one. She stretches and goes out to get her morning paper, the *San Jose Mercury News*, from the lawn.

Norman, working on his car, waves to Heidi. He is a purchaser and planner for a major company, and lives at home with his parents. His car is his pride and joy, the ultimate object of his salary and his stock options. It is in mint condition and "fully loaded." It has an expensive stereo—complete with cassette, CD and DVD players. The latter device is installed in the dashboard, fully retractable when not in use. When it slides out he can and does watch movies while he's driving. He also has a play station hooked up to the DVD player so he can play games. In the center console Norman keeps some cigarettes, some CDs, a bit of candy and some cologne. In the door tray he has a remote for his DVD and stereo players and a cell phone. In spite of San Jose's low crime rate, he has a spray can of Mace as well. He enjoys using the TV to entertain his friends, or to amuse himself while he commutes. He recognizes that it is a bit addicting, though. He told Heidi that he once took his friend to the mall to buy something, but rather than accompanying him inside, Norman stayed in his car and played games on his play station. Yet he considers himself a bit old-fashioned. He doesn't use a PalmPilot, but instead writes actual notes with pen and paper to keep himself organized throughout his day.

Norman and Heidi hold different values, but they are friends. Norman is Chinese—a fact Heidi hardly thinks about. His father is from Taiwan. Norman's "auntie" Lily, a friend of his parents', drives by. Lily is an interior designer. She considers herself an amateur ethnographer, as she studies her clients and gets to know their passions. Do they love the American Southwest? Have they traveled to India? Do they love to ski? She helps them decorate their homes with the appropriate "artifacts." She understands that objects can transport her clients to their "dream space" where they can relax, heal and be whole. Her clients, often overstressed managers, need to have an environment that can bring a bit of serenity into their lives on demand. Lately, being Chinese has actually added to her skill set, since feng shui is all the rage. Her network of fellow designers often discuss their problems. They must transform industrial park offices and suburban homes, however sterile and uninspiring on the outside, by providing interiors that have "a Silicon Valley look." The image is ultramodern with bold colors and chrome, softened by artifacts drawn from global traditions and tailored to the expectations and experiences of each client.

The Silicon Valley aesthetic is like a photo mosaic, a composite image consisting of many tiny photographs—such as a portrait of Abraham Lincoln made from hundreds of tiny Civil War photographs. While the broad outline conveys an impression of high technology and California living, the details of life reflect snapshots of people who come from everywhere, from Bangalore to Berkeley.

PART I

A Technological Place

CHAPTER 1

Culture Version 1.x

A Technological Community

Why It Matters

When California's Santa Clara County was labeled "Silicon Valley" in the 1970s, the region was transformed in the public imagination. But much of the mythic characterization of the region as a brave new world is hyperbole. Although the Valley is home and showcase for the latest in high-technology innovation, its denizens do not live lives radically different from those of their urban American counterparts. There are distinct social and economic classes. The institutions of the community—schools, hospitals, mayoral offices—are not so different from those in Sacramento or San Diego. People eat, sleep, work, and play in patterns familiar to many Americans.

Yet the region experiences forces that will significantly shape the future elsewhere in America, and the world. Technological devices from e-mail servers to telephones make it possible—even easy—for people to form dense interconnections in local networks, as well as in wider global affiliations. Technology suffuses daily life, the economy, and even the very language of Silicon Valley. Like the belled sheep at the fore of the flock, Silicon Valley is a bellwether beast, pursuing the newest technologies on the drawing board and in the hand. Its specialized economic history, once based on fruit agriculture and now built around high-technology production, has drawn people from around the world. The community's cultural complexity makes it an illustration of postmodern life. The heterogeneity of classes, ethnicities, national cultures, self-identified subcultures, and organizational cultures makes it difficult to assign individuals to any particular category or to assume that anyone shares your cultural premises. Artifacts and behaviors may derive from Midwestern homeliness, Californian counterculture, or from any number of sources from around the Pacific Rim and beyond. Europeans find the Valley European, while South Asians have reproduced bits of Indian life. Midwestern Americans find it both familiar and alien. Cultural interactions are inherently ambiguous; certainty in the cultural identification of oneself or others is illusory.

The things that make Silicon Valley distinctive—its technological saturation and complex range of identities—are not merely interesting cultural artifacts in themselves. They are significant because both the pervasiveness of technology and identity diversity are coming to define the emerging global culture. By studying the nature of the bellwether sheep, we may understand the consequences of technological saturation and cultural complexity for the rest of the flock. With this in mind, I have deliberately identified Silicon Valley as a natural experimental laboratory.

Silicon Valley is not the only place where either technological saturation or cultural complexity are dominant factors in defining culture. Indeed, if it were the unique repository of those features, it would be irrelevant to our understanding of any other culture. But "silicon places," whose economies are increasingly dominated by high-technology industries, are replicating around the globe, from Austin to Bangalore, while Manhattan, Chicago, and London are home to wide-ranging cultural diversity. Beyond these dramatic examples, even smaller communities feel the exponential growth of consumer technologies and the increasing opportunity to encounter people different from themselves. These people are also subject to the forces that so obviously shape culture in Silicon Valley. Silicon Valley has enthusiastically embraced technology and cultural complexity, making it a prime location for the anthropological study of what happens in any technologically saturated community. We can learn from its experiences.

This book is an anthropological foray into an emerging global landscape. The production of technology dominates the region and is attracting people from around the world, reshaping cultural identities. Silicon Valley has been studied by economists, urban planners, sociologists, business theorists, and historians. They examine their own particular slice of social reality, be it the structure of networked global business practices or the struggle of the underclass in the showcase region of late capitalism. Journalists capture the story of the day, often highlighting the rich, the famous, and the exotica of Silicon Valley.[1]

---

[1]Annalee Saxenian, an urban planner, examines the region for its connections to organizations, particularly as it compares to Boston's Route 128 and the role of Indian and

Anthropology is concerned with mundanity—the details of daily life, and what the small actions and interactions teach us about the human condition. The sites for that exploration and the tools for uncovering behavior differ widely. In the United States, anthropologists have been trained in a particularly broad disciplinary worldview, adding insights from biological anthropology and archaeology to direct observations of cultural life. This provides an interesting lens through which any time or place can be viewed. In addition to classical ethnographic inquiry— that is, observing and listening to living people in their own environments—we have additional conceptual tools. From biological anthropologists we learn to think about the processes leading to human variation and evolution that can be broadly defined as "change through time." We are, in the end, animals—but animals who manipulate our own environments and organize ourselves to adapt to the world around us using the ideas and artifacts shaped by our cultures. Archaeologists have taught us that sweeping cultural changes show a pattern when viewed over time, and that the tiniest objects we use reveal much about our behavior. Our words and actions tell stories, but so do our artifacts. A fragment of porcelain fired in Asia can tell a tale of international migration and trade, and illuminate the daily routine of a person who might have lived and died in obscurity, under the historical radar screen that tends to register only the prominent. These perspectives force cultural anthropologists to ask questions about the smallest details of daily life and then link them to ever changing larger forces. Hence, this anthropological consideration of Silicon Valley focuses on ordinary people, living lives filled with the minutia of daily activity, surrounded by material objects and cultural ideas. Evidenced in those small objects and behavioral impulses are larger evolutionary forces, vast historic changes that drive us to re-create our cultures, often without even knowing it.

Social anthropology, a once-British tradition now practiced around

Chinese entrepreneurs in the region (1985, 1994, 1999). Sociologist and urban planner Manuel Castells examines Silicon Valley as a global technopole, a twenty-first-century industrial complex (1996, 2000; Castells and Hall 1994). These works focus on the role of the entrepreneur. In contrast, sociologists Dennis Hayes (1989) and Karen Hossfeld (1988) examine life among the less privileged workers in Silicon Valley. Jean Deitz Sexton's *Silicon Valley Inventing the Future*, Po Bronson's *Nudist on the Late Shift* (1999), and Paulina Barsook's critical political commentary *Cyberselfish* (2000) are distinct examples of more journalistic examinations of life in the Valley.

the English-speaking world, teaches us that how we organize ourselves into groups, and how we support those organizational forms with beliefs, is a vital clue in unraveling human behavior. Silicon Valley people organize their lives around networks, family, and work organizations. These organizing principles are part of the distinctive culture that defines the region. As an anthropologist, I must explore those aspects of social life.

I am primarily a cultural anthropologist, and so I focus my attention on the role of culture as I find it in its natural setting, in the "field" (Lindholm 2001: 12). The idea of culture is one of anthropology's greatest gifts to social philosophy. It refers to "everything that human beings have created and transmitted socially across time and space" (van der Elst and Bohannan 1999: 32). Anthropologists are used to employing the term "culture" in its broadest sense, as in: "Human beings adapt to their environment using culture." In the past, the term also referred to the social entities that were presumed to share the same "creations" and "socially transmitted" ideas. We continue to refer, rather imprecisely, to the "Navajo culture," or the "American culture," a practice that tends to make us ignore the important contextual differences between Navajos in Window Rock and Navajos in Los Angeles.

It is unfortunate that this concept was first conceptualized as *kultur*, a noun, rather than a verb, as that is misleading.[2] Culture is the operating system that shapes our cognitive and behavioral processes, the "'conceptual structures' that create the central reality of a people" (D'Andrade 1984: 115). However, defining the scope of the "people" that create culture is problematic, since culture acts at many levels of social organization. "Creations" and "social transmissions" take place within the family, the network, the community, the region, and the nation. Culture "happens" across national boundaries at a global level in McDonald's restaurants, airports, and cubicles around the world. Yet all the people in a single family, or a single nation, share behaviors only in the most gen-

[2]Arjun Appadurai makes a similar argument when he suggests that culture is best used as an adjective, "cultural," in describing other aspects of life. He suggests that the "idea of culture as difference" best defines the use of the concept. Hence *cultural* differences can be detected ethnographically, even when cultures may be complex and fragmented (1996: 12–14). However, he restricts his definition of culture to only those aspects of social life linked to identity, a definition that excludes the material realm—"administrative arrangements, economic pressures, biological constraints, and so forth"—in explaining human behavior (Kuper 1999: 246).

eral way—demonstrating patterns, but not absolute uniformity. In studying culture it is important to look for the patterns—the footprints of commonality—while also documenting the variation within the patterns. Silicon Valley does not "have" a single uniform culture—although patterns do emerge—but it contains practices from many cultural variants in endless combinations, creating something altogether singular. This book lays out some of the cultural patterns that have been teased out of peoples' words, artifacts, actions, and interactions.

By looking at cultural processes at the community level, this study joins the many case studies of complex communities, from Hong Kong (Evans and Tam 1997) to Pittsfield, Massachusetts (Nash 1989). Specifically, I use Silicon Valley as a case study to reveal the experiences and consequences of technological saturation. This information has a special relevance to people who are connected with such "silicon" communities, both ordinary citizens and policy makers. Silicon Valley is also a natural laboratory for cultural complexity, containing a diverse array of interacting identities. Thus, while this study focuses specifically on Silicon Valley, it has wider implications for understanding the more general processes of living with digital technology and intense cultural diversity.

Silicon Valley also provides us with a mirror in which we can look at ourselves and examine our own choices. Some communities actively seek to duplicate Silicon Valley's apparent success, or at least those features that they believe will lead to prosperity, making political decisions that encourage industry, create private-public partnerships, and aggressively promote technical and infrastructural "progress." Corporate and public organizations enact policies less visible than acts of Congress, but perhaps not less profound in their effect. Individuals also embrace technologies for many purposes and results, reinforcing existing values and shaping new ones. Understanding the social life of Silicon Valley people allows all of us to reflect on the choices we make—both inside and outside Silicon Valley.

Digging Up Stories

This book is based on material from the Silicon Valley Cultures Project, a fifteen-year exploration of work, family, technology, and identity that began in 1991, conducted by Charles Darrah, James M. Freeman, and myself (English-Lueck et al. 2000). Because the project has extended over a substantial amount of time, it has described life in Silicon Valley during economic downturns as well as upswings, contractions and expansions. It spans a critical period in the community's history as it struggled to define itself and convey that identity to the rest of the world.

The project itself has constantly been redefined, sometimes reflecting small-scale efforts and at other times embracing major research undertakings. These ethnographic operations have been unified by a common thread: all have concentrated on the details of everyday life and ordinary people while at the same time noting the broader implications for "the big picture" of culture change.

The Silicon Valley Cultures Project comprises several different studies, some designed to capture breadth—such as our study of work, identity, and community, partially funded by the National Science Foundation. We chose a cross-section of 175 people to see how life was experienced from different classes, genders, sexual orientations, birth cultures, ethnicities, and professions to capture a slice of regional life. We interviewed temporary workers, administrative assistants, janitors, and assembly workers. We talked to various kinds of engineers and to the people who managed them. We talked to people who worked in start-ups and in established corporations. We visited national laboratories, such as the Stanford Linear Accelerator Center, and small machine shops, speaking with researchers and machinists. We sought out mothers who worked as engineers and mothers who worked as parents. We talked to retired people and college students. We visited working people in gyms, restaurants, churches, and schools. We interviewed the people shaping the images of Silicon Valley: marketers and media workers. We walked through the doors of Adobe Systems, Advanced Technical Staffing, Apple Computers, @Home Networks, Cisco, Detente Technologies, Hewlett-Packard, Phoenix, Peoplesoft, and Xilinx. We talked to outposters—people sent to absorb the culture of Silicon Valley, from Ericsson and from what was then Daimler-Benz. We queried nonprofit staffers from ARIS, the Community Action Center, and the Institute for the Future, and career and temp workers in city governments from San Mateo to Fremont (English-Lueck 2000a).

While we had done some comparative research that touched on information technology work—English-Lueck did research in China and Hong Kong on high-tech work organizations and identity (1997) and Freeman continued to conduct transnational research with the Silicon Valley Vietnamese and their familial networks in Vietnam (1989,

1996)—we did not yet have a broader picture of different "silicon places." So in collaboration with Andrea Saveri at the Institute for the Future, two of us—English-Lueck and Darrah—explored the global connections of the "silicon network." Collecting interviews in Bangalore, India; Dublin, Ireland; and the Taipei-Hsinchu corridor of Taiwan, we broadened our understanding of what it meant to live and work in a community saturated by technology. These places are gaining recognition for their technological work. We found that workers in each culture used their cultural knowledge and social networks to increase their competitiveness. Their social and national identities were among their tools. Their devices—telephones, mobile telephones, and internet connections—allowed them to form a new culture deeply intertwined with technologies. We also interviewed global workers in Silicon Valley who were part of this network of connections. With this final insight, made possible by interviews with sixty additional people, we began to understand what it meant to live in a technologically saturated global community.

Other projects investigated specific topics in more depth. Some studies were small and highly focused. For example, we collected interviews with vocational educators about the "skills" imparted in Silicon Valley, augmenting Darrah's research on work practice and the way people conceptualize and talk about skills (1994; 1995; 1996a). Interviews and observations with Joint Venture: Silicon Valley and the Tech Museum of Innovation gave us valuable insights about the cultural assumptions of organizations that shape community images.

Other in-depth projects were more ambitious. We knew, from our broad studies, that work and family were intertwined in Silicon Valley, and that technological devices played a significant role. So, with the aid of a grant from the Alfred P. Sloan Foundation, we conducted approximately twenty-five hundred hours of observation, shadowing twelve middle-class, dual-career families in Silicon Valley.[3] Each family came from a different cultural background, did different work, and had children ranging from two weeks to college age. We shadowed the children at home, in their classrooms, at play, and while they helped their parents with work. We shadowed parents working at work and working at par-

[3]The ethnographic field study of dual-career Silicon Valley households was completed in 2000, but only the preliminary analyses were available while this book was being written. A separate book will contain a more complete analysis of the interaction of family and work life in the region.

enting for hundreds of hours per family over the course of a year. We also documented two additional families for a shorter duration to explore particular cultural characteristics in Silicon Valley working families. We recorded each time that e-mail was consulted—for some once per day, for others twice per hour. Every child's game was observed and noted. Phone calls, trips to the doctor, school events, and business meetings, all were grist for the ethnographers' mill. We participated in the lives of the families and observed when ethnic or professional culture made a difference. This study provided additional depth to the insights we had gained from other observations and interviews about work, identity, and community in Silicon Valley.

While each project employed a unique set of tools, the overall tool kit remained constant. As anthropologists, we used anthropological methods. We recorded what people said in interviews, discussions, and "official" public documents. We made extensive use of the ethnographic interview; guided by an outline of topics for inquiry, we listened to responses and probed for details, rules, and stories. If we asked about identity, we were not satisfied with a simple "I am a Northern Californian," but hounded the person for specific definitions, including stories about times they enacted that identity. We probed to test the boundaries of identity—just when and to whom do they identify themselves as "Northern Californian"? We did not rely on words alone, but made detailed observations of people in their work and home environments—documenting the artifacts they used and asking them to explain the meaning of those artifacts.

In this "archaeology of the cubicle," material objects were mapped and discussed. The ethnographer was given a tour of each informant's domain of cubicle, lab, or assembly floor. As an item of living archaeology, each artifact was significant. What was that pile on the floor, and how did it differ from the one on the table? Who is in that picture? Why are there three calendars, the large one on the wall with no writing, one in the computer, and the Day-Timer? What information is put on each? Who has the right to add meetings to your schedule on the networked computer? Tell me about that cartoon on the evolution of power, showing a set of footprints beginning with bare feet, leading to men's dress shoes, and ending with a woman's high heels. The office artifacts that a person surrounded herself with might reveal more of her strong sense of "Irishness" or "womanhood" than her answers to direct questions about culture. A question about a photo of New Zealand disclosed the location of a cherished child. Snapshots of high Sierra lakes spoke of dreams, as did a picture frame, sold with a mass-produced photograph of a young girl, which a co-worker had picked up for Beth because it looked like the child she might have had.

We conducted observations—systematically watching and noting activities and interactions. We also used participant-observation, using our own experiences in the cultures to learn their rules and constraints. We immersed ourselves in the lives of the people we studied and in the community as a whole. Our own lives and families became part of the data.

Tom Fricke, an anthropologist who has studied people in both Nepal and the American heartland, notes that when we study an unfamiliar culture, we struggle to find the familiar. However, when "we work close to home," anthropologists must struggle again, this time to "look for distance" (2001: 24). In the course of my anthropological travels, I had done both kinds of research—with "distant others" in China and Suriname, and with California's holistic healers in familiar neighborhoods. I discovered, in the course of my fieldwork in Silicon Valley and elsewhere, that I could not always predict where I would feel "at home" and when I would encounter exotica. In studying Silicon Valley, I had to consciously strive to find familiarity in the landscape of technically saturated corporate capitalism, a cultural context alien to me. At the same time, insights from my work in Taiwan and Ireland, as well as my own prior experiences of living in radically different cultures, gave me emotional distance from the people I studied in Silicon Valley, who were my intimate peers. This mental yoga became even more strenuous as the journalistic world discovered "anthropologists studying Silicon Valley." I became part of the media machine that defines Silicon Valley, even as I struggled to question my own assumptions and collect information on its cultures. Doing ethnographic fieldwork in one's own backyard is an exercise fraught with irony. This is only appropriate, as irony is an attitude much appreciated among the people we study. Talking to an anthropologist gives Silicon Valley people an opportunity to reflect on their own cultural practices, often leading them to a chagrined reappraisal of a particular choice or incident. In turn, I cannot buy a pager for my daughter, or "try on" an interesting cultural practice borrowed from Japan, without realizing that I have become part of the culture I am studying.

Finally, reflecting our own intellectual interests as well as the fascination the region attaches to the future, we consciously sought to anticipate Silicon Valley's future while documenting the "ethnographic present," the slice of time in which we made our observations. We used a technique for uncovering unspoken assumptions and values by asking people to talk about their visions of the future for the region (see English-Lueck 1997; Textor 1985, 1995). Our major intellectual partner for considering implications for the future was the Institute for the Future (IFTF), a nonprofit think tank in Menlo Park, with whom we share a common interest in culture change. With them we investigated the role of emerging technologies in the household and at work, using ethnographic interviews and observations to illuminate the patterns that emerged from large-scale national surveys, anticipating the social conditions of the near future (see Darrah, English-Lueck, and Saveri 1997).

An Ethnographer's Sketch

Ironically, although Silicon Valley is the capital of "virtual reality," its geographical setting is all important (see map 1). It lies on the Pacific Rim, connected to Asia, Mexico, and Latin America. It is a deeply American community, currying favor with Washington, D.C., to support high-technology industries, recruiting from America's universities and businesses, and selling to an American market. It is an American culture as well—praising capitalism and celebrating individual merit.

Silicon Valley is part of the urban West, only a short air jaunt on commuter planes to Los Angeles, Portland, and Seattle. Linkages to Hollywood produce the special effects so loved by movie audiences. Connections to Intel's facilities in Portland and Microsoft in Seattle dominate high-tech businesses of the Cascades. Even more important to Silicon Valley urbanites, it is only a short drive to the ski areas of the Sierras, the redwoods of Northern California, and the beaches of the Pacific. Mobility between cities and a romanticized attachment to "nature" and "things Asian" are cultural features of the "New West."

Silicon Valley partakes of California's diverse social and physical environment. California is far from homogenous; it contains deserts, temperate rain forests, mountains, and plainlike valleys. The Bay Area in which the Valley is nested is likewise diverse. There are wetlands, mountains, and coastal areas, and the southern arm of the San Francisco Bay points toward the Valley. Europeans express dismay when they arrive in the Valley. Expecting a picturesque, alpine arc, they are instead greeted by two sets of haze-dimmed hills separated by a plain. To the southwest are the Santa Cruz Mountains covered with oak trees and conifers, and to the northeast the mostly brown, grass-covered Diablo Range, with such romantic names as Mt. Misery and Burnt Hills. Occasionally broken by low hills, the heart of the Silicon Valley is a flat plain a dozen miles across.

Geographically, environmentally, and socially, Silicon Valley is part of the Bay Area. Located on the inland coast, its climate is "rarely very hot or cold," an asset always commented upon by visitors (McCormack 2000: 230). The natural beauty and mild climate are "assets" to be cherished (Joint Venture: Silicon Valley Network 1998: 23). Locals express a desire to preserve open spaces, and, like other residents of the Bay Area, are likely to try to contain the previous decades of sprawl by encouraging "infill" or "recycling"—building in already developed areas (Lewis and Neiman 2000: 10; Joint Venture: Silicon Valley Network 1998: 30). From 1992 to 1998, two hundred thousand jobs were created in the Valley, but only thirty-eight hundred new residences were built (Joint Venture: Silicon Valley Network 1998: 24). This created tremendous pressure to use the land to develop new housing and business structures. Northern California culture dictates that residents are not like "those Southern Californians," who do not appreciate natural beauty and who live in nightmares of endless strip malls and suburbs. So sections of the mountains that border the region have been set aside from development. Yet, despite the admiration and affection they express for the natural world, Silicon Valley people live most of their lives in a human-made world of prefabricated tilt-up buildings, elaborate consumer spaces, and residential housing. The freeways, so crowded that they are sometimes referred to locally as "parking lots," carry the 17 percent of the high-tech workforce that commutes from neighboring counties (Joint Venture: Silicon Valley Network 1999: 12). Sidewalks, where they exist at all, are curiously underused.

People dominate the environment in which Silicon Valley residents live. In 1900 there were a mere two million people in California. There are now thirty-four million people, but they are not evenly distributed (Public Policy Institute of California 2000a). California's mountain ranges and rural areas contain fewer than half a million people, in an area roughly the size of Pennsylvania. Crowded into its coastlands are some of the world's great metropolitan areas: the Los Angeles metropolitan area contains seventeen million souls; the metropolitan San Francisco Bay area has more than seven million (ibid.; Malson and Martindale 2000). Silicon Valley itself, as reckoned by the Silicon Valley Manufacturing Group, has a population of 2,461,356, with 806,281 households (1999).

Silicon Valley is a prototypical high-tech community. Thousands of high-tech companies, not including a vast informal economy consisting of late-night software developers and corporate piece workers, buttress a variety of high-tech industries. Olivia, a civil servant in one of the local municipalities, describes the region's economic reliance on technology as a difference of scale rather than kind. She notes that people from all over the world have boasted to her about their latest industrial park, and "with some modesty [she] had to find some way to say when it was [her] turn that we have 6,200 high-tech companies in Silicon Valley and 350,000 high-tech jobs. I mean the whole valley is just immersed in this."

No one central agency or technology has a monopoly. Companies design computers; write the necessary software; design, build, and test computer network and telephone systems; design the semiconductors used in computing; and track down the human genome and convert the information to software. Apple, Cisco, eBay, Intel, Genentech, and NASA Ames make their homes in Silicon Valley. Thirty-seven percent of Silicon Valley's gross regional product comes from high-tech industries (DeVol 1999: 55). A British engineer, Stan, notes that "nobody else in the world comes close to the range and diversity of companies. And the complexity of their interactions is, I think, the single most unique characteristic about the region. Then, of course, you bring all these people into the region to fill all of these jobs."

Silicon Valley's population is sustained by a high-tech regional economy, but one that is vulnerable to recession (ibid.: 90). While robust in times of growth, such an economy is also susceptible to downturns. In July 1992, the economy of the United States was faltering, and locally, unemployment in Santa Clara County was 7.4 percent. The local workforce was beleaguered by military closures, a sustainedly high cost of living, layoffs, and reorganizations. In our earliest round of interviews about the future of the region, a lurking feeling of suspicion and insecurity dogged the informants' scenarios. However, by 1995 the situation was changing. High-tech industries were turning around. Myriad new software applications, microprocessor breakthroughs, and the birth of the electronic networking industry heralded a boom. By December 1997 the local unemployment rate was 2.3 percent (San Jose Mercury News 1998a: 12A). By October 2000 it was 1.7 percent (Steen 2000: 1C). The boom in the high-tech sector was felt broadly, as was the subsequent "softening." The technologically dominated Nasdaq stock index fell, and unemployment was back at 5.9 percent by September 2001. Layoffs were being measured in the tens of thousands (California Employment Development Department 2001a,b).

Silicon Valley has been dominated by successive waves of different industries drawing different people at different times. Joseph Schumpeter talks about economic cycles of "creative destruction" in which one industry booms, then levels out, only to be replaced by another. Silicon Valley illustrates that pattern, surfing successive waves of success based on defense contracting, integrated circuit design, personal computing, and internet software and hardware development (Henton 2000: 46–47).

This pattern has left behind a legacy of industrial diversity—although all are related to technology. Five percent of the industry is based on bioscience, 5 percent on aerospace and defense, 15 percent on semiconductors, 16 percent on software (including internet applications), 18 percent on innovation services that incubate start-ups, another 18 percent on professional services (ranging from printing services to attorneys), and 23 percent on computers and communications (ibid.: 51). Until the 1980s, financial and legal services were located in nearby San Francisco. Since that time, these services have moved south (Kvamme 2000: 72). Significantly, companies colonized these niches inventively, combining computing with bioscience to create bioinformatics, or moving from desktop publishing software to web design. Individuals likewise moved between industries, positioning themselves for the next wave.

Silicon Valley has a mythic history too, which accentuates its link with high-technology industries. Frederick Terman's seminal Depression-era electrical engineering department at Stanford, and the resulting creation of the Stanford Industrial Park after World War II, are part of the basic mythology of the region's coupling of technical innovation and entrepreneurial savvy. The legendary "garage" in which William Hewlett and David Packard created an innovative audio oscillator for Walt Disney's *Fantasia* in 1938 is now a state historic landmark. The defense contracts of World War II and the Cold War supported technological innovation at numerous research and development sites in the region.

In 1956, William Shockley coinvented the transistor at Bell Labs. He moved back to Palo Alto to found Shockley Transistor Company. His abrasive and autocratic style soon alienated the locally famous "traitorous eight," who left Shockley to join Fairchild Camera and Instrument, which later spun off Fairchild Semiconductor. That company, in turn, begat fifty new companies between 1959 and 1979, including Advanced Micro Devices (AMD), Intel, and National Semiconductor. In a 1969 conference on semiconductors, 95 percent of the attendees—then almost all male European-Americans—had worked for Fairchild at some point in their lives (Saxenian 1994: 30).

During the 1950s and 1960s, research facilities sprang up like mushrooms. General Electric Tempo, IBM, ITT, Lockheed, Philco-Ford, NASA's Ames Research Center, SRI (Stanford Research Institute), Sylvania, Westinghouse, and Xerox all developed facilities in the region (see Saxenian 1985: 24; Rogers and Larson 1984: 44; Winner 1992: 40). SLAC, the Stanford Linear Accelerator Center, was founded in 1966, attracting a rich mix of physicists from around the world (Winslow 1995: 3).

In 1971 journalist Don Hoefler coined the phrase "Silicon Valley," making the region into a symbol that has been replicated around the globe. Both nationally and internationally, there are numerous new technologically identified communities. Portland, Oregon (Silicon Forest); Austin, Texas (Silicon Hills, or Silicon Prairie); Albuquerque (Silicon Mesa); Ottawa (Silicon Valley North); Bangalore, India (Silicon Plateau); Taiwan (Silicon Island); Livingston, Scotland (Silicon Bog); and portions of Auckland, New Zealand (Silicon Bay) are examples.

Although the region's reputation for high-technology industry was already well established, it was the repeat performance of the garage scenario by Steve Jobs and Steve Wozniak, the founders of Apple Computers, that solidified the idea that innovators with entrepreneurial shrewdness could strike it rich. The personal computer revolution tripled the local employment base during the last five years of the 1970s (Park 1996: 162). And while the metaphor of the Gold Rush is used repeatedly to emphasize the Valley's financial promise, it is not strictly accurate. Rather than being based on mineral wealth, the prosperity of Silicon Valley is derived from its dense networks of skilled, mobile, and "diverse" professional workers, locally referred to as the region's "value-added." It is assumed that merely being in Silicon Valley, surrounded by different cultures and all sorts of technical experts, makes the people living there more productive and efficient. The real mother lode is thus the social capital embodied in Silicon Valley's knowledge workers.

The thousands of high-tech workplaces of Silicon Valley are themselves intricately intertwined. For example, rising sales at Cisco, the leading networking equipment company, caused a ripple of prosperity through a chain of suppliers including Solectron, a circuit board supplier; LSI Logic, a vendor of programmable semiconductors; Col-Kyl, tool distributors; Schulze Manufacturing, the manufacturer of Solectron's boards; Hadco/Zycon, a bare circuit boards supplier; Trend Plastics; Markovits and Fox metal recyclers; ECI Painting, the company that painted Schultze's cases; and Economic Packaging cardboard box makers (Thurm and May 1997: 1A). Workers at these companies need food, entertainment, and space.

Office space was quickly filled, as was rental housing. The rental occupancy rate rose to 99.5 percent, the highest in the United States, fueling the highest rents in the nation. The housing market increased at a rate of 15.5 percent, triple the national rate (Perkins 1998: 1A). Since 1992, one unit of housing has been built for every 5.5 jobs created, intensifying the housing shortage (Joint Venture: Silicon Valley Network 2001).

New employment niches begged for experienced workers. High-tech leaders pleaded with the Immigration and Naturalization Service and Congress for increases in H–1B specialty job visas, in order to import skilled labor from India, China, Canada, the Philippines, Taiwan, Korea, Japan, the United Kingdom, Pakistan, and Russia. People flowed into and through Silicon Valley to partake of this feast of work.

Californians use the technologies produced in Silicon Valley. While 72 percent use computers at home, work, or school nationwide, 76 percent of Californians and 82 percent of Bay Area people use these devices (Public Policy Institute of California 2000b). Sixty percent of Californians report they use e-mail to stay in contact with friends and relatives (Baldassare 2000a: 35). This pattern is mirrored by internet use—60 percent of Americans, 65 percent of Californians, and 72 percent of Bay Area people use the internet. Educated and affluent Bay Area people, regardless of ethnicity, use the internet at a rate of 82 percent (Public Policy Institute of California 2000b). The Bay Area has the highest internet use rate in California: more than half of the people use the internet to make purchases, and 57 percent use it to get work-related information (Baldassare 2000b: 33).

The primacy of technology is evident in the region's landmarks. More than any other single entity, with the possible exception of the Tech Museum of Innovation, Fry's Electronics is identified with the Silicon Valley life. Foreign dignitaries brought to Silicon Valley are taken to Fry's. Andrew, a journalist, noted in his interview: "[Silicon Valley] is about the way people are going to communicate, interact, gain information, pass along information. . . . It's about innovation, creativity. It's about diversity. . . . Diversity in types of businesses that are here, the types of people that are here. Silicon Valley is really like the future. The future is being created here. If I were to show them what it's all about, I think I'd go to Fry's!" This string of former grocery stores had "reinvented" themselves to become, quite literally, temples of technology, selling the "three C's—computers, components and convenience foods" (Langberg and Slonaker 1997: 1A).

Each Fry's store is decorated with a different theme—giant integrated circuits, crashed flying saucers, or Mayan or Egyptian temples. Fry's has been called "a living Silicon Valley Smithsonian" (Cassidy 1997: 1E). At one branch opening, the original Apple prototype, along with the Varian brothers' Klystron tube (which gave birth to the microwave industry), were on exhibition. Each store displays a vast expanse of consumer choices, from M&Ms to motherboards. Fry's is also a place where Silicon Valley workers can see the products they have developed being sold in the real world. Violeta, a long-time engineer, talked of taking her family for weekend excursions to Fry's. While children played games on the display computers or browsed among the CDs, parents grouped in technical corners to discuss the nuances of a part with other customers. These shoppers, once strangers, are instantly linked by their common passion for technology and their disdain for the "ignorant" salespeople. Going from one end of an aisle to the other, one may hear six different languages. It is impossible to exit the store without walking the "impulse buy" gauntlet—sugary treats and gizmos demanding attention from consumers as they wait in long checkout lines. A visit to Fry's reveals the underlying premises of the local culture—the importance of work, family, technology, and cultural complexity.

Use of technology is linked to work, the lodestone of Silicon Valley life. In our fieldwork, if we asked about technologies, we ended up hearing about work. If we asked about family, we heard about work. Work is a center of discourse. Work matters and workplaces matter.

Work is used to explain why a child needs a computer—to prepare her for the world of work. It makes the presence of a home office necessary. Bill comments how, back in 1995, he had thought that having a computer at home—in order to work there—would help him get ahead, but it turned out to be an illusion. Soon everyone had a computer at home, and Bill needed it just to keep up. Programs "borrowed" from work reside on "home" computers. Barbara calls her company's voice-mail system to retrieve messages in the predawn hours of the morning. Jeff does "work-work" at home, finishing tasks for his official employer, as well as "working" on improving his marketable skills and "working" on his far-flung social networks.

The presence of work organizations is strong. In spite of the overall number of high-tech companies—counted variously as seven to eight thousand—the large organizations, such as Cisco, Hewlett-Packard, Lockheed, or Apple, were key players in the community and in people's work lives. High-tech work is distinctive. Much of it is knowledge work—receiving, manipulating, and passing along bits of information through an interdependent network, often transnational, of fellow workers. The work involves creativity—often a great deal of it—and is not easily made rote. It is not easy to tell by observation if someone is working. Creative work is notoriously hard to monitor, so people make a great show of working. Theater is a part of work life, as individuals and teams demonstrate their dedication through working long hours and continuing discussions of work problems. The goal is "enhanced performance," a mission that seeps out of the workplace into other areas of people's lives. It isn't just a store-bought Halloween costume, it is one designed and hand-stitched by a working mother. That science project is not just another school event, but a chance to have your eight-year-old demonstrate the use of lasers, coincidentally reflecting the father's job. Each commute trip becomes an opportunity to create a more efficient route—to shave off a few minutes by driving "smarter."

Work enters people's lives in curious ways. We learned about the ubiquitous use in corporate life of the Myers-Briggs Inventory, a personality assessment tool. This test types people by their cognitive and relational styles. Gregory told us: "I have used it and I do use it. I find myself in meetings, in bars, anywhere I am willing to explain to someone the entire system to the best of my ability, in whatever time I have, to create a common language. . . . Say you are dating someone new, so who is he?

How are you getting along? . . . Well, here's his personality type. Here is my personality type. Here is where they click and here is where there are going to be differences." I observed other instances of the infiltration of work into the social fabric—when two men meet at the park, with only a passing acquaintance through their children, and share their work histories until they find a common link.

This exploration of potential relationships is reminiscent of an example from classical anthropology. Imagine a Tuareg warrior and his entourage riding across the Western Sahara desert of North Africa when they meet another group. First the men find out if the "stranger" is religious. Does he follow Islamic law, or is he "someone, who, not fearing God, may attack" them? Then they talk about the camps where they recently slept, indirectly establishing if, and how, they are related. Once the parties have identified each other, and established that they are not supposed to be enemies, they can exchange information about resources. Their compatibility is determined by the kind and degree of their relationship (Youssouf et al., 1976). The men in the Cupertino bar comparing Myers-Briggs types or two near-strangers meeting with their children in a park swapping job histories are not so different.

Work has also shaped the demographic profile of the region. At the beginning of the 1990s, people—mostly the less educated and the working poor—were leaving California for other areas of the United States faster than they were entering. By the turn of the twenty-first century, this emigration had leveled off. However, during that same decade, the well educated, the wealthy, and the foreign-born were coming to California (Johnson 2000: 1–6). California has become a showcase for diversity—50 percent of its residents are non-Hispanic Euro-Americans, 30 percent are Hispanic, 11 percent Asian or Pacific Islander, 7 percent African-American, and 0.06 percent Native American—and a full quarter of the population is foreign-born (Public Policy Institute of California 2000a). In addition to immigration, the diversity stems in part from the high birth rate of interracial and interethnic couples. In 1997, this amounted to 14 percent of the total births. Most of these children were born to native-born Californians; immigrants were less likely to have mixed-ethnicity children (Tafoya 2000: 4).

Americans are "Southerners" or "Californians," and they understand that each regional affiliation brings a separate experience of food, relational style, and cultural expectations. For many, the Bay Area itself defines an identity; it is a land of liberal thoughts, multicultural restaurants, and invent-it-as-you-go-along traditions. For some, particularly those who came to work in high-tech industries, the culture is a "Silicon Valley Culture." This identity, in which work defines worth, is based on producing technology, and embracing a fast pace and open attitude. In theory, it embodies the ultimate expression of personal achievement beyond the restrictions of one's birth. People revel in tales of bootstrapping—that is, social mobility. Tom, a highly placed general manager, describes how each of his colleagues comes from among the working classes of many different countries, but, in a parable of meritocracy, each is now in Silicon Valley playing for high stakes. People assert that opportunity is not affected by national origin, class, or gender. However, differences are detectable. The culture of opportunity looks different to a janitor, an admin, an engineer, and a high-tech executive.

While the details vary from city to city, the Bay Area mirrors California's general demographic profile. Silicon Valley is nearly evenly divided between California locals and newcomers. The Valley is like a giant village with two moieties—complementary tribal subdivisions—cross-cutting the region. One set contains locals—locally born people whose families may have lived in the region for generations. They have a set of experiences and perceptions of the region that are quite distinct from those of the people who have come as migrants, whether from Iowa or India. Newcomers come to *the Silicon Valley* with a specific image in mind. More than 20 percent of the people of Santa Clara County are foreign-born, and an additional 25 percent are from another state; slightly more than half of the residents are native-born Californians, some of whom presumably are from outside the Bay Area (Lyon 1998: 5). As more people came to Silicon Valley to participate in the high-tech economy and lifestyle, what our informants called "Silicon Valley Culture" became identified with a distinct way of life—"a technical place," as Cherice, a receptionist and massage therapist, put it. Many old-time residents are only dimly aware of its existence; others see it as a raft of job opportunities that allows them to stay afloat amid the rising cost of living. Migrants from the American heartland bring with them an image of open-ended opportunity and technical glamour, even as they go through Californian culture shock.

Much of the necessary expertise and support for Silicon Valley industry is drawn from a global pool of talent. Our earliest interviewees noted the dramatic transformation of the Valley from apricots to advanced technology. With mixed feelings long-time residents noted the post-1965 influx of new Asian and Hispanic immigrants into an already diverse community. Some embraced the changes, while others found the new ethnic landscape unsettling. Agriculture had attracted earlier waves of international and internal migrants, as Portuguese, Japanese, Mexican, and Filipino laborers came and went, or came and stayed (Matthews 1976; Ignoffo 1991). After 1965, a new wave of students and professionals came—some to stay, often going back and forth to their original homeland—bringing family and creating a mélange of communities. Vietnamese refugees and a continuous influx of Mexican and Central American immigrants added to that diversity.

Contemporary Silicon Valley high-tech industry attracts its own diverse workforce. A typical team at a local company included engineers from Bangladesh, Canada, China, Ethiopia, India, Iran, Japan, Korea, the Philippines, Taiwan, Vietnam, and the United States (Lewis 1993: 22A). Within the workforce, ancestral culture, nationality, and "racial" categories shaped and continue to shape a hierarchy. Asian immigrants themselves constitute a diverse group. Elite, highly educated Chinese, Hong Kong, South Asian, Japanese, and early Vietnamese immigrants are distinct from later waves of less educated Vietnamese, Cambodians, and South Asians. More than 28 percent of the systems analysts and chemical and electrical engineers were Asian, according to the 1990 census (Johnson 1992: 20A). European and Israeli engineers were recruited into positions with higher pay and status, while Mexicans and Fujianese Chinese—both legal and undocumented—worked alongside less educated native Californians in the manufacturing sector. Nearly 72 percent of all operatives and laborers were members of minority groups (Siegal 1990: 2). The majority of people working in the trades were Hispanic (Johnson 1992: 20A). Between the Mission period of the nineteenth century and the 1990s, the Native American sector grew from a small community of Ohlone peoples to a wide variety of urban Indians, including a small but growing cohort of American Indian scientists and technologists. The ethnic and cultural diversity of the area was felt in many areas of civic life, including politics.

California's sociopolitical landscape is complicated. Ethnicity, class, and gender are all factors that can be seen in the demographics of democracy. While Euro-Americans tend not to view their ethnicity as socially or politically relevant, other ethnic groups see political participation as an avenue for pursuing social issues (Cain, Citrin, and Wong 2000: 29). With the exception of particular Asian groups who view anything resembling socialism with a jaundiced eye, ethnic groups supported, and influenced, the Democratic party platform. In spite of the passage of a series of voter initiatives that appeared to oppose diversity—such as Proposition 209, which aimed to eliminate ethnic and gender preferences in California's affirmative action programs—surveys have suggested that most voters favor diversity in the workplace, higher education, and public life (ibid.: 5, 42). Interestingly, while Californians prefer to live in areas in which their ethnicity is in the majority, they also expressed discomfort with living in a homogeneous neighborhood or encountering attitudes of intolerance (ibid.: viii, 5).

In Silicon Valley ethnic culture matters more to some than to others. To some it is the source of identity and social connection; to others it is a nearly invisible backdrop in which their own individual concerns are more important to their perceptions. Not surprisingly, the foreign-born identify with their country of origin. As expected, many Americans express the deeply imbedded notion that culture is a function of genetic and national heritage. For minorities particularly, this is the cornerstone of identity. Hispanic culture is linked to food, family interaction, life goals, and the stuff of thought itself. To be unable to see the world through an additional Spanish language lens is to not be Hispanic. For others, national heritage is a distant fact—the arrival of Norwegian ancestors, some generations back, to a dimly remembered Midwestern town has little to do with one's personal identity. Occasionally, during an interview, an epiphany occurred, as when Leah realized that her Mennonite upbringing may have helped to make her the hardworking soul she is today.

The effects of living in the middle of such a complex social environment enter people's work and home lives in many ways. Tom, a senior manager and engineer, reflects on how diversity affects his own family:

Our daughter . . . has had a series of boyfriends. She was going with an Israeli whose family was from Israel and came here with a company. . . . Her next boyfriend was sort of a quintessential Palo Alto WASP-type kid—father's a doctor. . . . Her next boyfriend . . . was Vietnamese. Her next one was kind of a more normal American image of a Midwestern guy, and her current boyfriend, who may become a son-in-law, is Chinese. I kid her all the time: She has a couple of Mediterranean countries, Africa and South America to fill out her portfolio of boyfriends.

In his workplace, gay and lesbian concerns add to Tom's sense that everybody is not the same. Dramatic class differences, layers of ethnic and national migrations, corporate cultures, professional identities, and personal passions—such as a pursuit of natural healing and spiritual evolution—make Tom's community even more complex. Tom reflects on the local character of his church: "When we were first there, they had decided they were going to declare themselves a nuclear free zone. We also used to have, every other Sunday when we first came, a representative from . . . Nicaraguan or Honduran rebels giving sermons or participating in the service."

Like the rest of the Bay Area, Silicon Valley is overwhelmingly allied with the Democratic Party. After the 2000 election, only one state assembly seat was held by a Republican; the other offices—Senate, Congress, and state senate—were held by Democrats. Yet political life in Silicon Valley is qualitatively different from life in the rest of the Bay Area. The power of high-tech industry underlies political life there at every level. In the 1990s, Silicon Valley began to be felt as a political force in the nation, as local interests lobbied for H–1B visas, educational reforms, and Y2K preparedness (Kvamme 2000: 74–75).

Partnerships between local governments, labor advocates, and industries are nurtured, since they serve economic development. The Joint Venture: Silicon Valley Network, a government/business partnership explicitly designed to reinvent the region to be more favorable to business, points to a postmodern reincarnation of the company town (English-Lueck 2000b). A company town is a historical, largely American, phenomenon. When mining and logging companies moved into frontier areas that lacked infrastructure, the companies provided it by building churches, schools, and political structures designed to make life better for the owner company. In Silicon Valley, the marriage of government to business in such fields as education and health care is viewed as a significant "social innovation" (Henton 2000: 56). Such government-business partnerships have streamlined local permit processes, created a "unified building code," and established "smart permitting," in which local government interacts with applicants over the internet (ibid.: 57). Smart Valley, an initiative of Joint Venture, "led by technology industry business leaders" then "acted as a catalyst in connecting the region's schools, local government, and community to the internet." Doug Henton, founder and president of Collaborative Economics, notes that mayors and local government officials "reshaped local government services to meet industry cluster needs" (ibid.: 57).

A decade ago when I went to the San Jose State University library, I found a telling sign. Scribbled into the dust jacket of a dusty tome on the technocracy movement was written: "The movement is not dead, but lives here." Historically, this Depression-era social movement advocated the redesign of society at the hands of the technical elite. They sought to eliminate inefficiency and re-engineer global cultural practices. Although overstated, the unknown scribbler's observation appears to be reflected in local politics.

Among the key regional political issues yet to be confronted is the gap between rich and poor. Overall, California's income gap is increasing. Not only are the incomes of the affluent rising, but, in addition, the middle and lowest incomes are not responding to economic upturns. Underpinning this trend is an increasing "rising returns to skill" (Public Policy Institute of California 1999: 1). Educated people with concomitant skills are earning more than ever, while less skilled and less educated workers earn less. Unlike that of the rest of the United States, California's poverty profile is distinctive, for its many undereducated immigrant heads of households drag down the lower end (Johnson and Tafoya 2000: 3–9; Public Policy Institute of California 1999: 1–2).

Locally, this effect is both masked and exacerbated. Unlike the rest of California, the wealth of the lowest sector in the region is increasing (Thurm 1996: 1A, 25A). However, this economic fact masks the increased effort needed by the working class to meet the rising cost of living in Silicon Valley. Strategies for survival include roommates and multiple-family and multiple-worker households, long commutes to distant but less expensive housing markets, and multiple jobs. Food banks are used by working families that spend 70 percent of their income on housing (Bailey 1998: 1A, 16A). Only 58 percent of renters in the Bay Area believe that they may ever own a house (Baldassare 2000a: 34). The middle class is squeezed. Machinists described to us the four-hour daily commute to Tracy, sixty miles away, the nearest community in which housing is affordable. Jack, a vocational instructor, voiced the following concern:

The quality of life here has gone down. . . . When I first got married I was making less than five bucks an hour. I was probably making two or three bucks, and I thought that was really stylin'. When I got promoted to making about five bucks an hour in '68, I could live comfortably on that. My first apartment was eighty bucks a month, I mean, come on. You can't even come close to that anymore. It's just that everybody has to work, and you don't even have a good quality of life anymore. It's just so expensive.

Middle-class professionals live in group houses to be able to afford a nice view and a large kitchen. The volatility of the high-tech economy results in a constant sense of insecurity. According to the 2000 census, the median income in Santa Clara County was nearly sixty thousand dollars, contrasting with thirty-seven thousand dollars nationwide (U.S. Census 2001). However, our interviews with people revealed that the wealth was viewed as ephemeral. Wealth may exist only on paper, bound up in stock options or half-million-dollar mortgages. Reorganizations, project closures, and corporate mergers threaten individual financial stability.

Our informants saw money as a way of keeping track of relative status—nothing startling there. More interesting was the way in which money was recast instrumentally. It was viewed repeatedly as a way of sustaining what was truly important, supporting a comfortable lifestyle that enabled people to engage in creative work, not an end in itself. Of course, those with enough money to be comfortable can afford to engage in other obsessions—a high median income disguises the reality of living at the lower range.

Inequality also appears in the status of various workers. The ecosystem of the Silicon Valley worker is occupied by many various species—consultants, contractors, national temp agency workers, and temps from specialized firms. There is a parallel with Hong Kong's division of labor into *cheung-kung* (permanent), *cheung-saan-kung* (regular or long-term casuals), and *saan-lung* (short-term casuals) (see Kao and Ng 1992: 183). Each of these strata is treated differently by their employers and each has a distinct strategy for acquiring and keeping positions. Often, when civic leaders in Silicon Valley fret about workforce shortages, they are not referring to the entire workforce, but are primarily concerned with the technical and managerial elite, the equivalent of the cheung-kung.

The nature of the workforce, and its concomitant inequality, are gendered. Where technical prowess is the key to status, gender inequality is inevitable. Nationally, only 19 percent of the science, engineering, and technology workforce are women (Commission on the Advancement of Women and Minorities in Science, Engineering and Technology Development 2000), 20 percent in the Valley region (Collaborative Economics

2001: 5). There is a running joke that, not so long ago, university engineering departments were built without women's restrooms. Now women are granted a sanitary sanctuary, but they are still vastly outnumbered. In the 1998–99 academic year, 18 percent of San Jose State University's graduating engineers—the work horses of the region—were women, while 82 percent of the credentialed teachers were female (Institutional Planning and Academic Resources 2000). Ninety-one percent of the region's receptionists and 97 percent of the preschool teachers are women (Collaborative Economics 2001: 23). Compare teachers' earning power and social status with those of their technical counterparts, and you begin to see the shape of the gendered inequality in the region.

Differences in power play out in the details of daily life. The way people use their technology acts as a signpost pointing toward the role of power. To whom do people talk, and how? When is it preferable to leave an e-mail, or a voice-mail message at 2:00 P.M., rather than to communicate with "sneaker mail," a face-to-face conversation with a coworker down the hall? How do people learn that just because an e-mail *can* be sent to the corporate president, it does not mean the message *should* be sent? Power relationships are expressed through technology within households as well. Some people are expected to leave their pagers on at all times, while other members of the family can turn the cell phone off. Children can be tethered to parental control by a pager. Power means having access to others, while limiting access to oneself. This principle is even more visible in the workplace. "Efficiency" means that the lives of the more powerful are made more convenient by the labor of others.

Technological devices allow people to shape their work to fit immediate needs. Planning ahead gives way to "just-in-time" solutions. A patchwork of e-mails, faxes, and phone messages, pulled together by underlings working in crisis mode, rescues a manager who had not prepared a presentation to his client in advance. Technology makes it possible to develop one's schedule ad hoc as the day's events unfold. Why plan when you have a cell phone? Such devices enable just-in-time approaches that maximize personal flexibility. However, while some people benefit by that flexibility, others pay a price. When a nanny must change her own family plans to accommodate a shift in the parents' work day, she is the object, not the initiator, of the change. Power matters.

Family life, too, is molded by the use of technology. Different kinds of families use the same devices quite differently, sometimes drawing family

into face-to-face interactions around the VCR karaoke machine, sometimes flinging them into different orbits with electronic pager/cell phone/
voice-mail tethers. Gifts also create links. What is given to whom? Erika
told us of the computer she gave to her mother in hopes that she would
take up e-mail and regain her position as central communicator of the
family. Once, Erika's mother had connected the geographically spread
siblings into a network with stories told on the telephone. Now Erika
and her siblings e-mail each other, and mom is left out of the loop. Perhaps the computer would restore her to the family.

Technology is used to create and demonstrate family. Giving one's
wife a cell phone for her birthday is a demonstration of virtue by a caring
spouse, though it also sends a subtle message about the need to stay accessible and maintain the order of the family. Making oneself available
by pager is a display of parental conscientiousness. One mother told us
an impassioned story of the day her son ripped his pants at school. Because she could pick up the message he left on the family answering machine, she was able to find out about the crisis and bring him another pair
of pants, saving him from humiliation. This was not a trivial story to that
parent. That day the device was a tool for saving family face.

I cannot overstate the importance of family in Silicon Valley. While
there are certainly single people and childless couples, they, too, are connected to parents, siblings, nieces, and nephews. The importance and
value of family were clear in the answers to our questions about holidays
and celebrations. Great thought is put into gifts for key family members
and friends. An appropriate gift bespeaks intimacy, and a sense that in
spite of her chaotic schedule and reduced face-to-face contact, the busy
daughter does indeed remember that her father loves classic cinema and
her mother loves wildflowers.

Families can be sanctuaries from work, or themselves become integrated into the world of productivity. Other institutions—clubs,
churches, and schools—can likewise be treated as refuges or as platforms
for making useful and productive connections. People are linked in networks that occasionally overlap with institutions, but it is the networks
that count. Networks are important conduits of knowledge and reposito-
ries of trust. Dense networks, reinforced by many different kinds of ties,
are one of the defining sociological characteristics of the community (Castilla et al. 2000: 219–20).

Family—sometimes parents and siblings, often spouse and children—

includes those in the household and beyond. Former spouses and step-children might be more distant, but are still part of family life. Fictive kin, close friends that became the children's "aunts" and "uncles," are included in family, particularly for those geographically far from blood kin. Circles of acquaintances are generated, divided into "friends of the heart" and "friends of the road." Some of those friends have connections going back to earliest childhood, others are classmates or former co-workers. Occasionally, contemporary coworkers break into the circle of intimacy and become "friends of the heart," particularly if they were colleagues in a past job as well. There might be multiple links among different areas of life. Jeff worked with Aaron at Apple and now works with him again at Adobe. Jeff has children, as does Aaron, and so they eat lunch together in the cafeteria and get together on weekends. They both like science fiction and music, as well as sharing a passion for software.

The power of networks in daily life is amplified by the extensive use of technology. As earlier generations had found with the telephone, computing and wireless communications devices reinforce rapid nongeographic connectivity, upsetting the traditional way of dividing social environments. E-mailing the person in the next cubicle as readily as one e-mails someone in India changes the relationship of face to place, so that even physically local relationships began to become abstracted. Geography weakens its domination over "closeness."

The relationships of people—within networks, as well as more formal social institutions—are ruled by pragmatism. People grant each other favors, and the reciprocal exchange of information and services is a glue that ties the community together as surely as the sharing of meat in a small-scale band society. Relationships are also transformed into products, and people "work" on their relationships, making them projects with goals. Working on one's "parenting," or on one's romantic relationships, demonstrates an approach toward family and education that mirrors the practice of an engineer in the workplace. Life is a series of projects.

Knowledge is required to work on these projects. A dense network of educational institutions allows people to consume knowledge at will (see Castells 1996: 5; deVol 1999: 45; Saxenian 1994). The familiar duo of Stanford University and the University of California at Berkeley are only the most well known institutions. Other private universities in the region include Santa Clara University, the University of San Francisco, Golden

Gate University, John F. Kennedy University, and St. Mary's College. The University of California also has a campus in nearby Santa Cruz, and State Universities in Hayward, San Francisco, and San Jose, the last also housing the National Hispanic University. Community colleges such as San Jose City, Evergreen, De Anza, Foothill, West Valley, Gavilan, and Mission prepare students for university admission, as well as offering job retraining or continuing education to other students. Continuing education is also available through extension and distance education programs. There are also dozens of vocational institutes. One can attend workshops on "How to Master Deadline Pressures," "Landing Big Contracts," and "Self-assessment for Career Planning." One can belong to organizations that offer networking opportunities as well as training, such as the American Business Women's Association or Forty Plus. Adult and continuing education programs offer courses from Adobe software applications and bookkeeping procedures to conversational Vietnamese and Kamatuurun Kali, a Filipino martial art. A constant stream of conventions and expositions, a well-developed network of in-house and contracting corporate trainers, and a host of specialized magazines and web sites offer even more specialized training.

Religion also reflects this pattern. While as elsewhere religion is experienced in institutions—mosques, churches, and temples—Silicon Valley people "work" on their spiritual growth, shopping for cultural options and "the best" practices. For example, Aaron's wife converted to Judaism, reviving her husband's natal religion, so the family joke goes, in order to ensure that he comes home at least one night a week, on Friday, for the Shabbat meal.

While Catholics are still most numerous, Silicon Valley is the home of a mosaic of beliefs ranging from fundamentalist Christian to radical gay paganism. In our interviews, people talk about the role spirituality plays in making life meaningful. Whether imbued with Buddhist, Christian, Hindu, Jewish, Moslem, or neo-pagan tones, linking oneself to the great mystery is a passionate pursuit for many. The networks in which they express their spirituality are central to people's existence. Whether it is a women's group, a Zen retreat, or a local Catholic church, such networks of meaning-makers are important in people's lives.

Silicon Valley's ethos merges pragmatism and the search for meaning. While people "use" other people as tools for advancement, they also are bound in a web of genuine affection and reciprocal exchanges. In Silicon Valley's moral landscape, people should *share* information and services, not *take* them. Silicon Valley people struggle with the contradictions of daily life. The late-capitalistic impulse to glorify mercantilism and materialism exists side by side with a romantic idealism that recognizes the nobility of discipline. Technological devices are not simply material devices but also symbols of discipline and order. Tales of the "lost weekend" spent struggling with the computer are reminiscent of an older generation's struggle with Latin. The skills are useful, often in an indirect way, but the real reward is in the mastery of a difficult task. Struggles with technology are perceived to have a moral component. Being able to "connect that stupid peripheral" or "upgrade that impossible operating system" are signs of character and persistence.

Even the production of technology is viewed as a moral mission, reflecting a technologically inspired vision of progress. This ethos is not limited to the "digirati"—the technical elite—but is also articulated by admins, machinists, and artists. The Valley offers many potential niches for workers to contribute to something that is "changing the world." While functionality and practicality are clearly paramount values, the production of technology takes on aspects of spirituality, from Apple's special "Evangelists," who make marketing their products into a zealous mission, to the subtle joys of "creativity" that so many people find central to their work experiences. Silicon Valley also views itself as a community in which cultural differences matter less than the technological "mission," in which the bigotries of "other places" are left behind. These romantic notions are also part of Silicon Valley.

The Double Helix

What is the effect of technology in the lives of Silicon Valley people who are both the producers and consumers of that technology? How does culture, or rather the interplay of many cultural identities, matter in this place? Silicon Valley's culture, like DNA, takes the form of a double helix. In living organisms, DNA consists of two polynucleotide chains, running in opposite directions, that are wound around a central axis. It controls the synthesis of specific organic products and is the "transforming factor" that marks one strain of organism from another. New culture is made in Silicon Valley as two strands of cultural life intertwine.

Cultural ideas and practices about technology form one strand. They are inexorably intertwined with a less obvious strand of cultural richness reflecting diverse interacting identities.

Technological presence is so dramatic in Silicon Valley that it is easy to overlook the second, more subtle, impact of cultural complexity. Yet that second strand is well recognized by some of those who live entwined in the double helix. Eugene, a retired mechanic and Asian-American church administrator, notes: "On the one hand you see the high tech, and sometimes that's where it stops, because that's all people tell you about. But you also see the other side of it. You see the humanity." Heidi, a young native of the area, understands this in her bones when she remarks, "I just kind of get this image of this area . . . when I think of Silicon Valley, I think of diverse technology, and diversity within people, so, I just think of diversity, but technology is kind of booming in the background."

Rachel, a business editor and journalist, voices the same conclusion, adding:

I really think that you can define [Silicon Valley] by technology, but . . . then you only hit one slice of this place and if you factor in the diversity you actually then begin to hit much more of a whole. [My friend] did this story which to me is like the ultimate Silicon Valley story . . . about the Santa Clara Cricket Team. And that to me was like the most perfect and amazing story. . . . These people changed the Santa Clara Cricket Club from this moribund, horrifying, "Will we ever win? We just play cricket kinda bad," to this cricket powerhouse which, you know, is doing all these different [community activities]. They were lured here by technology for the most part, which is why they came here. But in coming here they have changed this place.

But how do they change this place? The influence of cultural richness and identity diversity on the high-tech lifestyle is a central mystery to be investigated. What is the difference that culture makes?

To answer this question meant rethinking how culture functions in communities radically different from the ones studied by my anthropological predecessors. In a village town in Indonesia, culture is created through the complex interactions of interpersonal obligations, social appearances, and individuals acting out unstated and shifting rules and roles. In a stunning example of how the whole can be viewed through the part, the anthropologist Clifford Geertz analyzed a Balinese cock fight, unraveling how the activities of the event expressed deep cultural premises (1990: 113–21). He demonstrated how viewing the "public culture" gave the anthropologist a window into the underlying premises of that culture. I had to locate the Silicon Valley equivalent of such cock fights. Evidence of public civic culture can be found in the political, economic, and artistic arenas. Public ritual is expressed in Joint Venture: Silicon Valley Network forums and celebrations of the Tech Museum of Innovation. Yet it is the theater of everyday life that is intriguing—when a person buys a gift shirt at Sears, tracks down a thorny technical problem at work, and plans the day's logistical contacts while commuting. An event in public life need not be grandiose to embody cultural premises.

These mundane examples can be found in Fresno or Des Moines, not only in Silicon Valley. So what is different about Silicon Valley? Silicon Valley produces technology. Its denizens are predisposed to use high-tech devices, providing ample opportunity for anthropologists to study a culture in which "public" interaction happens in electronic spaces. Faxes, voice mails, telephone calls, pages, e-mails, and web-based communications together create a device-mediated "public space" rich in cultural premises. Important cultural work is conducted through the use of electronic communications technology—reshaping social roles, constructing meaning, creating tacit agreements about which parts of a person's birth culture will be emphasized and which aspects will be overshadowed by common concerns. As electronic public space increases in scope and importance, it shapes a society in which technology is deeply integrated into all of life, in both gross and subtle ways. Technology defines the interactions of the day, and shapes the metaphors of the larger civic sphere. What do we have to learn from such technologically saturated communities? What can Silicon Valley tell us about the ongoing creation of new culture, which anthropologists refer to as ethnogenesis?

Not only does the technology shape the actions of the individual and the community, but, in addition, the political economy of technology production has an additional, unanticipated effect. The expertise for technological production is not local, but taps into global sources. That expertise is imported by electronic messaging or by actual migration in and out of Silicon Valley. People with diverse identities interact, and create new identities. The resultant cultural complexity poses a dilemma. While the high-tech economy cannot thrive without worldwide connections and continued global interaction, there is a certain amount of resistance and ethnocentrism on the part of Santa Clara natives and among immigrants as well. Cultural differences have always posed a problem in human interaction.

Ethnocentrism refers to the feeling that one's own culture is the "center of what is reasonable and proper in life" (Brislin 2000: 44–45). Assuming that one's home culture defines the only proper way to act, however, is at the heart of many distasteful and even violent intercultural interactions. Colonial history abounds with snap judgments and heavy-handed interventions. Classical ethnocentrism emerges from a sense of certainty, a "gut feeling" that what is familiar is inherently proper and what is unfamiliar is somehow suspicious. However, in a complex society it gets much harder to distinguish between "us" and "them," as identities overlap and cultural practices are drawn from many sources. Ethnocentrism itself becomes a more subtle process, reflecting unease and uncertainty, rather than certainty. People in such circumstances cannot assume that the people they meet share the same cultural premises about work, family, time, honor, or fairness. They cannot necessarily predict that those premises are not shared, either. This ambiguity, the uncertainty of sharing common assumptions, is the "new ethnocentrism" (see Geertz 2000: 86, 224). Philosophers and psychological anthropologists endeavor to understand how a society can work when its members share shifting and varied premises.[4] What do people have to do to overcome this "new ethnocentrism" and create a meaningful plural community and a productive work space? Silicon Valley's historically generated cultural complexity provides a stage for viewing the consequences of identity diversity.

The saturation of technology into daily life generates new social traps—cultural problems in which short-term and immediate solutions inadvertently generate long-term problems. Information technologies are adopted to create efficient, asynchronous, global communications. It seems so simple and convenient to e-mail project instructions to a subcontractor in Ireland from home before breakfast. However, while using devices may overcome the immediate technical difficulties of crossing time and space, this use has long-term social consequences that may prove problematic. The interactive "public" communities that are cre-

[4]The philosopher Charles Taylor writes extensively on the role of common intersubjective meanings in creating a functional polity. He defines consensus as the "convergence of beliefs on certain basic matters" (1985: 36), and contemplates the implications of multiple centers for countries such as his own culturally plural Canada. This cultural definition of consensus does not mean that everyone in a society shares the same opinions, but that they share a "common understanding of symbols" that permits civic discourse (see Levine 1984: 68). His work has been influential to anthropologists such as Clifford Geertz.

ated by electronic media are distant, often conveying only partial or imperfect communications, and to maintain them requires a fair amount of invisible work in the form of frequent contact and redundant communications. Device-mediated communication initially masks cultural differences, creating the illusion that the people on the other end of the phone or keyboard are "just like you"—at least until people discover they have quite different ideas of what constitutes such culturally loaded concepts as "timely" or "responsible." Even finding the right person to talk to about a decision may become a time-consuming and culturally laborious task for a project manager in Dublin who needs to talk to a counterpart in the United States. Who is in charge? When she finally locates that person, should she be assertive, or deferential? How can she convey the right attitude over e-mail?

Pervasive technology also sets up another problem. When information is mobile—either because of the widespread use of mobile communications devices or easy access to stationary personal computers and telephones—how do people use the old environmental cues of workplace and household to divide the social realms of work and home? When a parent is generating human resources reports on a laptop while attending a child's sporting event, is that person acting as a worker or a parent? The enactment of social roles has been associated with place since the Industrial Revolution (see Nippert-Eng 1996: 19). Increasingly, people must invent new ways of managing their social selves.

High-tech work in general, and global work in particular, is deeply social. People must work together to exchange information, pass tasks to specialized workers, and learn their organizational culture. Information and communications devices facilitate the technical connections. But they also facilitate a wider net of connections to friends, family, and former coworkers, strengthening the salience of the person's overall network and weakening the power of the immediate employer.

The demand for specialized workers from all over the world has created a cultural complexity that makes it impossible to simply assume that everyone around you is just like you. Once again, there are unintended social consequences to this identity diversity. The short-term need to extract the technical elite from a global talent pool is solved by importing labor from India or Boston, but the importation results in greater cultural complexity. Some of the people with whom one is interacting may be housed in distant parts of the world. Other people who are culturally different may be at the next desk, or dropping their children off at one's children's school. This is a social state that is rife with potential cultural mistakes. A simple shopping expedition may require effort in developing cultural sensitivity.

In this book, I must describe a region where life is saturated by technology, and where identities are problematical on account of the complexity of cultural interactions. There are places in the world that share aspects of Silicon Valley's technological penetration and identity diversity. Each of these conditions also has ramifications for other kinds of communities, those that are not identified with technology, or which, at least on the surface, seem culturally less complex.

User Guide

As a cultural anthropologist, I am prey to a lifelong fascination with the details of daily life, and a predilection to see any culture as but one among many. I have been trained to look for ethnocentrism in myself, to constantly question my tendency to think of my home culture as the natural one. Tempocentrism—viewing my own time as the default setting for normality—can also lead me into error. I did not grow up in a reality dominated by hyperactive electronic activity. I must resist the temptation to assume that any culture, including any in Silicon Valley, is inherently flawed or favored, particularly while I am in the process of trying to understand it. That does not mean I cannot detect bigotries and contradictions, or note particularly creative cultural solutions to dilemmas, but my training inhibits me from ranting either in praise or condemnation. Yet, I too am thoroughly enmeshed in the Silicon Valley system. I live and teach in Silicon Valley. My children think of it as home. I grumble at the traffic and grimace at the inequalities, just as other Silicon Valley workers do. I am subject to occasional bouts of "technolust," going to Fry's Electronics just to gaze longingly at the latest digital device, a sure sign of my having "gone native." Yet my experiences in different cultures, and decades of exposure to the study of diverse cultures, help me place Silicon Valley in perspective. Hence, throughout the book, I draw parallels with, and distinctions from, other cultures.

Anthropology is a discipline rich in metaphors. In my particular kind of anthropology, we are keen to understand the context of a person's life, and we toy with the idea that "the metaphors by which people live and the worldviews to which they subscribe mediate the relationship between what one thinks about and how one thinks" (Shweder and Bourne 1984: 189). In the early days of the field, we labeled the gross characteristics of a culture with metaphors. Ruth Benedict extracted the imagery of Friedrich Nietzsche to label the Zuni "Apollonian," and psychoanalytic metaphors to highlight the "paranoid" aspects of Dobuan culture (1989). In the postmodern turbulence of the latter twentieth century, culture was viewed as "text," and "interpretive" analysis made the use of metaphors as a communicative tool a common practice (Geertz 2000: 16–17). In addition to looking at political economies and social organization, studying metaphors allows anthropologists to glimpse the underlying moral reasoning of the members of a culture.

The very name "Silicon Valley" is a technological metaphor. The people use a variety of metaphors, often drawn from high-tech life in day-to-day discourse. In this book, I add my own technological metaphors, invoking analogies in chapter titles and headings to set the stage for the content of each chapter.

The people of Silicon Valley do not form a single, undifferentiated entity. In our research, we listened to many voices in Silicon Valley, ranging from those of the technically elite "digirati" to those of the janitors who clean the school hallways. But this is not a book just about the hyper-affluent, or the clearly abused underclass, but about the vast middle ground in between. In portraying the experiences of such a range of people I was confronted with a challenge. There are many characters in the story of Silicon Valley, enough to make a classical Russian novel seem simple. The identities of those people must be kept anonymous. I wanted to keep true to the voices we heard, while not betraying any identities. Thus, pseudonyms were assigned to the various people who informed this study. But merely letting people "speak" in quotations carefully matched to their alter egos robs us of the richness of their experiences. People live in a world of artifacts, actions, and interactions that cannot always be represented in speech alone. To convey that richness, each of the subsequent chapters will begin with a vignette that illustrates the ideas discussed later in the chapter. These scenarios will take you through a day in Silicon Valley, beginning with the morning commute, and ending with the setting sun. Direct quotations in the vignettes come from real people in Silicon Valley and other "silicon places," but their identities are masked. Some characters in the vignettes are fictional composites whose behavior is based upon hundreds of observations and in-

terviews. The composite characters are created to enliven widely observed actions and to create fictional integrity in the scenarios. The detailed vignettes are themselves fictional composites. The words and
deeds in them were said and done, but by a number of different people.

Conceptually, two ideas dominate this book—technological saturation and identity diversity. Silicon Valley showcases changes in daily life
that come directly as a result of the pervasive use of technology. The region also embodies changes in demography, and highlights the complex
cultural interactions that accompany participation in a global high-tech
economy. Hence the book is divided into two parts.

The remainder of this book explores the intertwining strands of Silicon Valley, as the prototype of a community that is suffused both with
many technologies and many identities. The economic specialization of
the region has drawn people with great technological expertise to the
community in unparalleled density. Technology permeates everyday life
and provides the metaphors of community identity. The magnet of high-
tech work has created a new population influx. Historically steeped in
agrarian-based ethnic diversity, Silicon Valley has drawn different
populations from within and beyond the United States into its high-
technology economic engine. The array of cultures in the region fueling
the workforce ranges from Cambodian culinary entrepreneurs to Midwestern process engineers. International ties emerge not only from immigrants and economic sojourners but also from the social bonds that are
made and repeatedly reinforced through emerging electronic technologies. The region is not only a bellwether of technological research and
production but also a laboratory for the creation of a complex society
that contains diverse identities. Individual identities emerge, engage,
erode, and are re-created to produce a larger community of communities
in which people interact in schools, workplaces, and homes.

The two strands of technological saturation and identity diversity intertwine to produce many different choices in uses of technology, work
practices, community connections, and family relationships. One dominant pattern emerges from these choices—instrumentality. Instrumental
reasoning—the kind of reasoning that calculates the relationship of
means to ends—is integral to producing and using technology. Life is
managed. How does that reasoning affect the way in which we live as social and cultural beings? What happens when people make cultural identity itself into a tool, an instrument that is a means to an end? How do we

manage identity complexity in an increasingly global culture? Once anthropologists traveled to distant islands in the South Pacific to test the validity of established gender roles or socialization patterns. Today we recognize that life in Silicon Valley is a laboratory for the integration of consumer technology and transnational migration, reflecting larger American, and even transnational, cultural trends.

The first part of the book emphasizes the consequences of technological saturation. In "A Technological Place" I have considered the impact of technology on daily and community life in Silicon Valley. In the chapter you are now reading, "Culture Version 1.x: A Technological Community," I have introduced the outlines of life in Silicon Valley, both familiar and exotic. In Chapter 2, "Compressing: Using Digital Devices to Shape Space and Time," I look at what it means to be "technologically saturated" in everyday life, and how technology use affects the choices people make and the consequences, often unintended, of those choices. Here I discuss the significance of "work," a commonly used English word, but one that takes on a distinctive metaphorical meaning in Silicon Valley. Chapter 3, "Networking: Building Community in Silicon Valley," discusses the social organization and public life of Silicon Valley. Networks—a form of social organization that is facilitated by technology—dominate how people structure meaningful groups. Technological metaphors influence Valley language and create a distinct public culture. Community is "designed," "invented," "reinvented," and "refreshed." Civic activities include "NetDays," when high-tech volunteers install an infrastructure to bring the internet into public school classrooms, and the development and celebration of the Tech Museum of Innovation. I examine how the identification with technology is used by people within Silicon Valley, and in other technologically saturated regions, to create a "value-added" community.

The second part of the book is centered on the interaction of diverse identities. In "Trafficking in Complexity," I track the global movements of people that have shaped the cultural complexity of the region. In Chapter 4, "Input/Output: Emerging Global Culture," the evolution of Silicon Valley's global dimensions is revealed in detail, highlighting the complex mixture of ancestral, national, and corporate cultures that flow through the Valley. In it, I must unravel the role of culture and the function of identity. The fifth chapter, "Executing: Culture at Work and Home," explores how culture is viewed, identified, and used at work and at home. When is one's birth culture invoked? When is it avoided? Especially noteworthy are the strategies used by workers to "manage culture," although management of culture is not confined to the workplace. The limits and creative solutions families make to adapt to different ancestral and corporate cultural expectations reflect another way that culture is used. These choices are expressed in courtship, child-rearing, and interpersonal relationships. In "managing culture" Silicon Valley denizens engage in the ultimate metaphorical act of instrumental reasoning: turning cultural identity and cultural competence themselves into tools. Finally, in the last chapter, "Reformatting: Creating Useful Culture," we consider the coexistence of technological saturation and identity diversity, and the implications of combining these two forces. What are the challenges inherent in living in such a rich technological and cultural ecosystem, with so many choices and possible interactions? What are the tools people are creating to manage this complexity? How does it change how they organize their lives and relationships? How do the symbols, metaphors, and values of a community—so heavily identified by technology—shape civic life? Ultimately, what can we learn from the natural experiment known as Silicon Valley?

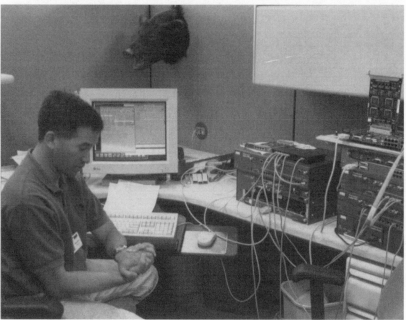

Top: In Silicon Valley, even the highway billboards emphasize technology.
Bottom: Cubicle land where work spaces are customized to facilitate work and
remind people of their other "lives." (Photographs courtesy of Thomas J. Boures)

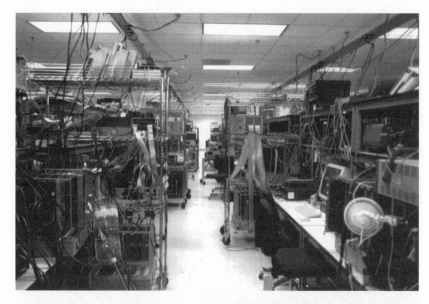

Top: Inside the world of high-tech work, networking hardware dominates the scene—an Asynchronous Transfer Mode (ATM) product development laboratory. *Bottom:* High-tech work involves collaboration. This software development team is having a team meeting in a manager's office. (Photographs courtesy of Thomas J. Boures)

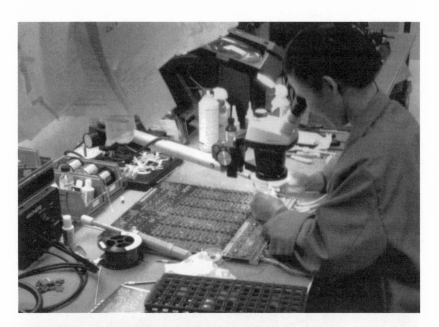

Top: An Asian woman uses a dissection microscope in a prototype development rework station. *Bottom:* The company provides dinner after hours during a "crunch time." (Photographs courtesy of Thomas J. Boures)

In addition to being a janitor in a high-tech workplace, this woman has other jobs as well. (Photograph courtesy of Thomas J. Boures)

CHAPTER 2

Compressing

Using Digital Devices to Shape Space and Time

Around the Water Cooler

Nearly every Silicon Valley workplace has a fueling station, containing at least one water cooler, with hot and cold water and a stash of tea and coffee. More elaborate stations have fresh-brewed coffee and chilled colas. It is midmorning in Silicon Valley. People are migrating toward the water cooler for that additional jolt of warm caffeine; those on flextime are starting their workday with a Starbucks refill. Jeff wanders along the distinctive carpet patterns designed to lead him to the upscale water cooler. There he meets Aaron. Both Jeff and Aaron are software engineers. Technology drips from their very pores. They greet each other and chat for a few minutes. Aaron gives Jeff a CD he had borrowed. Jeff has a stellar collection of music that circulates between his office and home. They chat about their wives, kids, rumors of reorganization, and the latest pages on their favorite developer web sites. After a short time they go back to their stations, pleasantly refreshed by human contact.

Elsewhere in the same company, in a nearly identical water cooler/caffeine haven, Sam greets Alex. Their talk also turns to technology, an enduring topic. Sam loves to tease Alex about the appearance of his office, more computer lab than private office. Alex refers to his many screen environments as virtual real estate. His third computer is a four-year-old Mac. He keeps it as his e-mail computer and for storage of old projects, a living archive of previous work. It is too old for doing any serious coding work: "I don't trust it much for that because it crashes all the time; on the other hand it has three screens and they're nice and big and one of them is color, so I tend to do my documentation work almost exclusively on this machine." Sam asks Alex how he ended up with so many monitors, privately wondering why he didn't jettison some of them at the last office move. Alex replies, "I initially started with just the one little color monitor. And then I got the big black and white screen that was just a hand-me-down from somebody who got it and decided he wanted to move to color, and I just needed it to put up text which it's great for . . . system editing a lot of programs. And then I got into doing some color work and I needed a [better] color monitor, so I just got the third one."

Sam had to ask why Alex got a PC after so many years working happily with the Mac. Alex chuckled. "I had the Unix machine and the Mac but the director of our division realized that most of the customers for our group are using Windows NT, and he simply decided he's going to just issue them to every engineer in his group. So one morning we came in and it was sitting on the desk. You put computers on engineers' desks and regardless of what they may have said about that before, they will start using it. Open it up—you can't resist. I was just amazed at the speed. It's the first PC I've used in . . . God, I don't know . . . ten years or more!"

Alex is doing cross-platform testing on three different platforms to track down problems, when software will run on one platform but not another. Alex tells Sam in frustration, "Those are ugly problems to solve. There's no good reason why something should work on one machine and not on another, but that often happens, and those are difficult problems to track down." So he really needs all those different machines to do his work properly. Hence the conversion of his office into a "lab."

In another water cooler area in yet another high-tech firm, the same conversation is echoed. Nathan and Marshall are discussing how they are adjusting to their new spaces. Nathan puzzles over the configuration of his space, his for the last month. "Frankly, the noise is an issue, and the fans, the heat would be an issue if the temperature control in the building was adequate and it wasn't too cold all the damn time." They both laugh, pointing to each other's sweatshirts. Nathan adds, "Space is an issue. . . . I need the desk surface, so I put all that stuff underneath which makes access to reconfiguring the network . . . an issue. Length of cables of things like video monitors, that's very high-frequency stuff. You just can't have a long cable on it, but that's an issue as far as flexibility—configuring." Like Alex's office, Nathan's cubicle looks like a computer lab. There is a Sun station that cost more than his college education. He doesn't use it if he can help it. He really doesn't like the Unix environment: "It's uncomfortable to use, the mouse is uncomfortable, the keyboard is nonergonomic, the software is bad, it's unreliable, it's tied in with the network, and I don't like the way the network and stuff is set up here. . . . The PC version is better, it's more reliable, and it doesn't crash and lose your work, the mouse motion is easier, it's better, a lot of reasons. . . . I'm using Visual C, so that takes up an entire screen, so I can still only get three or four files on here, and then lots of times you'll be editing a document at the same time so you bring up the editing application and that sort of lives on top, and then you've got your other stuff—like there's this MeetingMaker thing that helps you arrange meetings—and then you've got a lot of stuff that you get off the web, so you've got a web browser. I need four of these monitors all at once, it's just the way I like to work . . . just zooming between four applications . . . shuffling papers on the screen. It's annoying. . . . The main thing is the distraction, you're thinking of one thing, and what you need . . . you've got to rearrange three or four windows to get there." Nathan and Marshall commiserate about the need to juggle all these applications at once. Even when you are being left alone by your coworkers there are so many things to attend to, just doing the job.

At another water cooler station Stan and James are sipping hot tea. Both are from Britain and enjoy a few moments chatting in familiar dialects. James, an engineer, is working on a quirky compression algorithm. Data must be coded in a way that saves storage space and time, particularly when it will go over the internet. Files need to be grouped together in an archive, as a set of similar files, and then they can be compressed more efficiently. Stan teases him that his whole life is spent compressing, trying to find ever more efficient ways to make his work fit into his family. James counters with a "You too!"

Stan is the "techie" for his church, responsible for making sure computers are connected and telephones function. Stan also sets up and calibrates the sound system for the "contemporary Christian music" that is supposed to lure the hip, disenchanted "Silicon Steve" to the church. Stan tells James about the sound seminars he is running for folks on the church worship team. Stan's wife runs the congregation database. Thus he can compress his community and family life with his work passions.

Stan jokingly threatens to give James his jar of rocks, his favorite metaphor for juggling life's many tasks. He ran across this story in a training video "back on the East Coast." In this video, the guru "gets this jar out, and he says, 'Here is this empty jar, and I am going to put this stone in it.' So he gets these great big stones and he puts them in the jar. And maybe he can get four or five stones in it. And he says, 'Is the jar full?' And they all say, 'Yes, of course it is full.' So then he gets this bag of pebbles out, and he pours the pebbles in, and he says, 'Is the jar full now?' And the students work it out and they say, 'No.' And so he says, 'You're right.' So he gets out a bag of sand and pours that in and says, 'Is it full now?'" Stan pauses for dramatic tension and then continues his story: "And they are really clued in now, and they say, 'No, it is not full.' So he gets a jug of water and he pours that in, and of course it all fills up and he says, 'What is the message here? What is the point I am making?' And they all said, 'Well, the message is however full or however busy you appear to be, you can always fit something in.'" James nods, but Stan carries on: "And he said, 'No, that is not the message at all. The message is, if you don't get the big stones in at the beginning, you won't get them in afterwards.' The point being, make sure you have got the right big stones in your jar, in your life . . . before you start putting the small stuff in. Otherwise, you are not going to get the big stones in!" Stan adds that "once you fill your life up with garbage, with stuff that really doesn't need to be done, you are not going to get the important stuff in. . . . You have got to get the big stones in first!" Stan and James like each other partly because they have picked the same big stones—family, good citizenship, work, and the excitement of technology.

Stan tells James that the night before he had fought the good fight commuting to Fremont, and then had gone online to look up information on *Wallace and Gromit,* a current video favorite. This witty British claymation series features an inventor and his dog as the central characters. Stan typed in "Wallace" on a search engine and up popped a porn site. He rants to James, noting that the listing was right there, "the first, not like down the list somewhere, the top line item was some hot web site somewhere. . . . I don't know what you can do!" He wants his sons to grow up with the ideal of Christian chastity, but tells James one has to be "ever more vigilant."

Almost finished with his tea, Stan tells James about his home office. Stan has his "faithful Mac" connected to an ISDN line. Usually, however, he does his nighttime e-mail through his portable Think Pad, which is connected to a ricochet wireless modem. That way the messages are stored in the machine, very convenient for moving from place to place. He needed that device when he had his recent surgery and was bedbound for a week. He was able to stay in touch with his team while he was physically absent and thus not be overwhelmed by the volume of communications when he returned. He has a zip drive and a scanner, a cradle for his laptop, a color printer, and a very large high-resolution screen that can connect to his laptop. Behind these devices are all the power lines and the cabling they need. Stan is critical of his current setup and wonders if James can help him. It doesn't look bad, it is "all hidden, it is just a great mess back there. There are three power strips behind there. Ridiculous!" Someday he will have to do something about the "cobbiness," that unaesthetic quality of having cobbled together a graceless solution. James nods sagely, only partly paying attention, his mind pondering the process of compression. They finish their tea and walk back to their offices, ready to apply their expertise to the efficient merging of task and time.

Pivotal Technologies

When archaeologists excavate an artifact—perhaps an obsidian chipped-stone tool fragment—there are many ways to think about its significance. Archaeologists look at material, morphology, and function. The obsidian fragment can be examined as a made object, or as volcanic glass whose chemistry can be pinpointed to a particular lava flow. The artifact is a corner piece, so the archaeologists might be able to tell that it was shaped in a way that indicates that it was produced during its peak of fashion, maybe five hundred years ago. The edges might bear microscopic evidence of use, indicating contact with animal rather than plant tissue. They retrace the trade routes that brought that piece of volcanic glass to a region of the California coast that contains no volcanoes. The obsidian fragment embodies a complex production history: initially chipped from a core obsidian bomb, a teardrop-shaped chunk of solidified lava, hundreds of miles from the site where it was shaped into its final form. The archaeologists ponder the value of the object, rare in an area where the native stone, largely Franciscan and Monterey cherts, holds an inferior cutting edge. This tool had been carefully resharpened, and perhaps over generations had become an object of status. The obsidian tool is more than a mere object—it is the focus of a whole complex of human relationships. That tool is the stuff of culture.

In Silicon Valley, the tools of the trade—devices for computing and telephony—likewise can be looked at in many ways. They are tools of production without which new technologies could not be created. They are also tools for building and maintaining relationships, tying people together in a web of communications and connections.

Obligations are created as people help each other to use technology. A whole folklore has been created around technological devices. Devices have become status symbols and metaphors for life in the postmodern world. The use of the tools—be they telephones or PalmPilots—creates an "electronic water cooler," a social space in which people interact, construct identities, and exchange obligations.

Anthropologists are familiar with the pattern. An ethnohistoric indigenous Australian group from Northern Australia's Cape York Peninsula, the Yir Yiront, gathered plants, fished, and hunted, with domesticated dogs in tow. Central to their economy was the polished stone ax. It was a piece of capital equipment, used to produce firewood, construct huts, collect food such as honey, and to make ceremonial objects. Like the obsidian in Coastal California, it came from far away, four hundred miles to the south. A series of face-to-face exchanges between pairs of people took place, as they traded the stone ax heads for the barbed spines of the stingray that were used in spears. Only adult men made and could own the axes. Relationships were reinforced through the original trade. Webs of obligation were built up as younger people, especially women, borrowed the axes from the elderly men who owned them. Beneath the surface, however, the axes had a symbolic importance that was not so visible. Axes were symbols of masculinity. They were the totem, the clan emblem, of the Sunlit Cloud Iguana clan, an identifying symbol of the group. The stone ax was a pivotal technology, important to economic, social, and symbolic relationships.

The significance of this technology was highlighted when steel axes infiltrated the society. The dyadic person-to-person relationships built by trade and borrowing were undermined as European-Australians reinforced a social structure that relied on groups and leaders, a one-to-many relationship. Gender roles were subverted and the original network of trade was supplanted by the distribution of the new steel axes. Even the worldview of the Yir Yiront was challenged when the Head-to-the-East Corpse clan, associated with the Europeans, became identified with the new steel axes, causing a crisis as contradictions arose in the mythic foundation of society (Sharp 1952). Which clan "owned" the ax? That was no longer clear. Why should anyone favor the elderly men when they were no longer the source of the ax? Even women could own steel axes. What happened to the web of connections built by the trade of the stone ax? The lesson of the stone ax became an iconic story, reminding anthropologists to think about the complex role of technology as both tool and symbol.

Silicon Valley has its own complex of pivotal technologies, devices that structure the lives of people, acting as critical capital in production, shaping relationships between people, and providing a symbolic framework for their understanding of the world. Information technologies particularly—computing, telephony, and entertainment devices—are integrated into the daily world of Silicon Valley denizens. People have a unique relationship with these devices, for not only are they eager consumers but they are also the producers of these technologies. Their driving passion for technology makes them distinct from the people of other regions. Their technologies are their raison d'être and the economic prime mover, but the devices are also the chief production capital that enables the work itself.

People don't just own or use individual devices, but ecosystems of technologies located on their bodies, at various work sites, and at home. Pagers, faxes, cell phones, telephone answering systems, and computers are used together to serve the goals of individuals and families. Individual relationships are mediated through these devices. The use of technology is not trivial, but underpins important cultural work done by individual workers and families to create the networks they need to support their jobs (Darrah, English-Lueck, and Freeman 2000). People must adapt to an environment saturated with work and cultural differences. The devices, like the stone axes, are tools for creating relationships. People use their devices to manage the complex rhythms of high-technology work by granting communications access to some, to gain information and deflect access from others, and to gain time.

The need for trust and personal connections helps to drive the penetration of work into home life. The pivotal assumption that work is done at a workplace and family life is lived at home is much too simplistic for Silicon Valley life. Many forces, not the least of which is the technical capacity to work from home, have blurred the boundary between domains. The presence of a networked computer at home simultaneously grants an adult the privilege of working near family, and distracts that person's attention from the people in the home environment. The interpersonal concerns of an individual worker become the basis of discussion within the family and become part of family lore. The technologies also allow families to put old behaviors and relations into new contexts, using the communications skills honed at work to establish ties to far-flung kin.

As among the Yir Yiront, technology does not just play an economic role in defining work communities and families, but also a metaphorical, symbolic one. The technologies are also symbols of people's connection to a new cultural reality. As information technology allows households and communities to become places of production, it also alters the metaphors that people use to think of themselves. Families and communities, like upgraded software, can be "refreshed" or "reinvented." Families and communities themselves can then become a kind of product. The language of technology becomes the language by which life is discussed. The processes of technological production, with models of forecasting, management, efficiency, and technical problem-solving become the cultural processes by which life is lived. Just as the ax could serve as the foundation of Yir Yiront cosmology, information technologies become the chief metaphors by which Silicon Valley life can be understood.

In this chapter, several aspects of the role of technology beg to be explored. For example, technology saturates people's lives, but what drives that saturation? What choices are possible? What shapes those choices? Finally, what are the consequences of those choices? Some of the legacies of intense technological use affect culture—how people live. Others have an impact on identity—who people think they are. Both culture and identity are key ingredients in defining the Silicon Valley community.

Using Technology to Make Technology

In Silicon Valley, the creation and manufacturing of a product is an intensely social event. Whether parsing the human genome or designing a new internet router, high-technology work is networked activity. Information must be passed from worker to worker, each one taking the bit, transforming it, and passing it along. Members of research and development teams must ultimately link with webs of suppliers, producers, and customers. Beyond the individual, the workers in particular sectors function in tandem with traditional coalitions, such as the Silicon Valley Manufacturing Group (see Castells 1996: 191–95). Communication between members of a network is vital and ranges from joint decision-making to chatting about family to humanize and warm the tie.

Global networking, a feature of what is commonly called the new economy, is facilitated by an increasingly dense ecology of tools. Telephone conferencing services, Personal Digital Assistant programs, custom watches, and automated telephone operators that tell the caller the time zone of the receiver help people manage their international telephone calls. Asynchronous tools such as voice mail and e-mail make messaging convenient at any time. Cell phones allow calls to be made to Asia and the Pacific at seven at night and Europe at seven in the morning—during the commute—so that mother can come home to eat with the family.

People have clear strategies for juggling the many communications options. They use a variety of devices seamlessly, and often simultaneously. Both sender and receiver may have definite preferences for particular devices for particular purposes. Communications preferences cluster around cultural and functional divisions. For example:

—A worker leaves a voice-mail message for a difficult subordinate in the wee hours of the night. The logic is carefully articulated. At two in the morning one can leave a message stating a decision or a command without risking an energy-draining conversation. There can be no argument with voice mail.

—A programmer knows to send e-mails to fellow engineers but voice mails to people in marketing—that is just how they do things at his company. It is the local custom. Besides, engineers are plugged into their computers to do their work; the phone is a secondary tool. Mobile marketers live on their phones, and do not have constant access to computers.

—Schematics are sent by e-mail to an engineering teammate in Dublin to facilitate the communication of detail, but phone calls later ensure that the context for the design change is understood. One method records fine detail and creates a record; the other smoothes personal relationships and minimizes misunderstanding.

—Sending a faxed page with a signature requires a phone call to make sure the fax machine, known to be several offices away, is checked. The fax is semipublic property; the phone call can be directed to a particular individual. Otherwise, the fax might lie unobserved for hours or days; worse, it might be observed by the wrong person.

—A phone call from an upset quality assurance engineer quickly precipitates a face-to-face meeting. The emotion expressed in the voice clearly indicates that the situation must be handled delicately, with full access to facial expressions and body language.

—A scheduled video conference call between Bangalore and Silicon Valley reinforces the personal interactions that took place at a conference in January. The call builds on already formed relationships and once again links names to faces.

—An admin knows that the product lead gets from fifty to three hundred e-mails a day. She needs to draw the lead's attention to a time-critical item, so she goes in and places a Post-It note on the engineer's computer reminding her to check her e-mail for information on her latest travel advance. Workers jokingly call this method "sneaker mail."

—A municipal official keeps work and home separate by creating a barrier of physical distance—a common strategy. She commutes several hours a day to be able to maintain an affordable and distinct home life. During that commute she uses her cell phone to begin and end her management day. Her action has led to a "voice-mail organization" at city hall in which e-mail contact is minimized. A different official who did not commute and primarily communicated with e-mail would change the activities of the entire organization.

Each of these examples shows a different social context for communication. Within that context people choose different devices according to availability and preference.

The tools enable communications-savvy people to juggle the effects of absence and interruption. Elliot notes that his high-tech colleagues often face five-hundred e-mail messages upon returning from vacation—an overwhelming hurdle. One strategy for managing this flood of information is to access messages while gone. Sandra, an international operations manager, travels frequently in her work. She sighs, pointing to a map in her office, and notes:

Well, that's a big world! That's why I have a laptop instead of the desktop machine and that goes with me absolutely anytime I leave the States. Most of my trips are international, and this is the life line for communication. We've now got it set up so that it is fairly easy to dial right into the company and get past the firewall [that keeps out outsiders], so I can get access to servers, I can see our internal home page . . . with links that you can't see from the outside. Plus I can get all my e-mail and that kind of thing, so yeah, it's important absolutely when you travel. . . . It's all built in, lightweight. . . . They're making hardware today that is just amazing; you used to have to bring all kinds of different plug converters and power converters. I still need the plug adapter, but you don't have to have power converters, you can plug right in. They've got dual voltage built into these things. They get lighter, so they're pretty phenomenal, and the battery lasts longer than it used to. [She chuckles and adds:] So you can work on the plane and not have to power up.

The sheer quantity of communications means that devices need frequent attention. Determining the timing of access becomes a major part of the working day.

Rhythms of the Dance

In high-technology work the networks are maintained not only across space but also within time. Work is governed by the timing of activities, patterns that reflect "rhythms, cycles, beginnings, endings and transitions" (Barley 1988: 125). A delicate web of planning and scheduling allows people with diverse and dissimilar work rhythms to fit them together into an apparently seamless whole (ibid.: 128). Products are born, developed, and pushed to market in exceedingly short time frames. Noel, a software engineer, speaks of the luxury of the old days, when a product cycle might have taken five years. He works with routers, the devices used in electronic networks to direct data traffic to particular destination addresses. He notes that the devices he designs are increasingly complex, and that "you might find that a router that we build here is a lot more complex. . . . Certainly there is more code in there." Despite this increased complexity, the time frame for the product cycle is relatively short. Noel says: "A year and a quarter, year and a half; that is a pretty long product cycle here." That is eighteen months for a new product, with a novel design. He adds, "And if you are doing kind of an incremental product . . . like an interface card you put in your PC . . . for one of our routers, four- to six-month product cycles are very common." He then puts all this into context, by contrasting the current state of events with a much older cycle that included a period of less frenzied activity between projects. Noel states bluntly, "If you are a farmer, there is no winter. And there are basically four summers in a row. You have to hit them all, or the farm goes bust. That is . . . how it feels."

Each product cycle consists of a whirlwind of activity. Diverse players—computer scientists, various different engineers, technicians, machinists, prototypers, tech writers, marketers, and support staff—all work at different paces. Aaron likens the sensation to "dating too many people." Managers set and evaluate milestones to set a common pace.

Margaret, a tech writer, comments on how she fits into this time frame. She must interact with the developers of a product on the fly, describing the product as it is being developed. Over e-mail she and the technical staff will send and edit information literally every five minutes to get a "release note out the door within days." She spends time getting information on the product for the document and processing that document as her own "product." Her document gets a product number and is put on a schedule list. Customer support and manufacturing want to know what is going to be in the database. "What are the dates of the reviews? Cycles? How many pages is this going to be? What kind of paper is it going to be printed on? Is it going to be on a CD, is it going to be on the web? What's the title going to be? And then the simple things like, who's the writer, who's the editor? When is the freeze date for the content?" Once a freeze date is set, the technical product is not going to change substantially. After the freeze date, Margaret's work changes. During the communication-intensive aspect of her work, she lives on e-mail and voice mail. She communicates less and writes more. She has stopped gathering information, and now begins to produce the release note. The time period for her part of this process might be two to three days.

Marcia, a sales staffer who must make sure that the product is appropriately licensed for an end user, talks about her communication rhythms. She notes:

Mostly what happens throughout my day is to come in the morning and first thing I do is check e-mail and check voice mail and kind of react and respond to those things that are quick and easy. And then, usually, maybe something left over from the day before, a project. Usually our group is constantly being challenged from our field sales force with either voice mail, e-mail, very seldom a fax, and then [they're] just calling throughout the day. So we're dealing with the day to day stuff that is in the background. . . . There are a few goals and objectives that are due on a certain day, and so, depending on when something is due, you might be more available to just pick up the phone. Sometimes it might go to voice mail, but usually in the morning the very first thing you do is check everything to make sure that there are . . . no hot issues that need to be absolutely dealt with.

The daily rhythm of work is dictated by "fire fighting"—managing problems that must be solved immediately and are usually somebody else's fault—and squeezing in progress on one's own product, be it writing code, designing routers, or filing travel papers (see Perlow 1997: 87). Other people mesh their own intensively creative work, which is less amenable to scheduling, with the more routinized work of disparate co-workers. Jeff juggles his need to wander around "shaking lapels" to get the information he needs to start a new project with the "seat time" he must spend in meetings.

Compliance—use of the technology to conform to work rhythms—is a choice that can be made or subverted. The value of a mobile PDA (Personal Digital Assistant) still depends on how people use the technology. A manager distributed PalmPilots to her staff hoping to coordinate them more effectively, but the effort began to fall apart almost immediately. Only one person, David, fully utilized the device, while his coworkers fell steadily behind in digitalizing their appointments, events, and deadlines. The management dream of perfect coordination and organization was defeated by a lack of consistency in the staff members' adoption of the new technology. They resisted synchronizing their somewhat chaotic rhythms.

Some work is steady and predictable, rarely requiring extraordinary effort. Client-based work may necessitate a nine-to-five rhythm, or, if global, require a seven-to-eleven spread to include other continents' time zones. The final push at the end of a product cycle may fuel the notorious ninety-hour work weeks, followed by some days of collapse. Jennifer, who was an in-house corporate trainer, reflects on the joys of her predictable schedule, pointing out that "some of the groups around here . . . are ordering dinner in every night and sleeping in their offices."

While a start-up might require an intense pace, a more mature organization might foster a more predictable schedule. Indeed, Rennie, a Taiwanese engineer, was warned by his wife that he should either leave his start-up or expect to be left by her. He joined a large corporation, trading his intense and unpredictable start-up schedule for fewer and more predictable work hours, ultimately leading to a more stable family life.

Professional rhythms are blended with family and community obligations, resulting in ever more complex juggling (Hochschild 1997: 45). Flexible scheduling allows some people to work in mini-shifts. One person may start at 6:00 and work until the children wake at 7:00 or 8:30 A.M., then feed the children, take them to school, and drive to work after the commuter rush. She then works into the evening, confident in the knowledge that her spouse left his workplace at 5:30 so he could be back in time to take the kids to soccer and feed them dinner. She goes home to play with the children until bedtime, puts in another hour or two of work, and then it starts all over the next day. Another mother cannot separate work and home so neatly and must "multitask," calling to make the child's doctor appointments while checking the latest e-mail on the product release. Another mother nurses her baby while making a call to London to talk to a colleague.

The deluge of demands can be managed through a variety of strategies, some typically "Silicon Valley." When a high-tech organization finds its human and technical resources insufficient to meet the goals of a project, one solution is to outsource part of the job to an external contractor. Outsourcing may temporarily import critical expertise into the organization, or it might simply allow an organization to remove the overhead from noncentral production tasks, such as training or tech writing. When maintaining a staff of in-house trainers is too expensive, outside "training consultants" are hired. Prototypes of new devices are not made in-house but subcontracted—that is, outsourced—to a prototyper.

Everyday life in Silicon Valley becomes transformed into a series of projects, and the underlying logic of outsourcing extends beyond the workplace into family and civic life. As people push harder to meet the obligations of "work-work" and the commitments of "life-work," even changing the oil in the car oneself, or shining one's own shoes, becomes impossible. The same process of outsourcing is at the heart of how more affluent Silicon Valley families make life into work. Cliff outsources the car servicing, even though his happy memories of "fixing the car with Dad" make him want to fix it himself. Victoria has Amazon.com wrap and ship the gift she selected for her brother from the internet. She might have had a lurking suspicion that she should have wrapped it herself and sent a hand-written note, but there simply aren't "enough cycles to burn" to do everything herself.

Personal assistants and professional coaches colonize this niche and take care of the details of life that cannot be done in the few hours left over from work. Kerry calls herself a "pimp of time," who earns her living delivering chunks of time to her clients by doing a variety of activities. Such "extra selves" may be hired by individuals, but companies also may retain personal assistants who run errands for elite staff, releasing their time and energy for "work-work." Personal assistants may buy gifts or hire remodeling contractors as needed. In addition, housekeepers, gardeners, and nannies do the household work, releasing parental time to be recycled into more "work-work," networking, or working on recreational skills. Families are selective in which tasks to outsource. Lynn reflects on her choice not to have a nanny, whose values may not match her own, but instead hires cooks and cleaners to free up her time to be a "parent." Linda schedules weekend gardening as a joint family event to promote cohesion, but the difficult plants are left for professional gardeners.

Others choose diverse ways to reduce work or home obligations. Mothers who want to stay at home try to find work that can be done during the children's nap or school time. When Ada's son called her at work, she made communicative noises without listening to him while he tried to tell her he was hurt. When she realized what she had done, she quit her job. Joanna put her career on hold to stay home with her two children. She dreams of a career as an internet humorist, a female Dave Barry, and she is developing her writing ability in online communities. It's an electronic public space she can participate in while the children are sleeping. People shift and shape job and family demands until they find a way to make things work

Who Leads?

Andrew Feenberg, a philosopher of technology, notes that technology affects social status in two ways, by "conserving the hierarchy" and by subverting it (1991: 92). The people we observed and interviewed also used technology both to reinforce and to subvert old roles. Access and control are central to the use of these devices, which can be used either to control or to challenge traditional parental and gender roles. Rules about technology use are created to control family roles: "You must wear your pager," "You must carry your cell phone," "You must not use the computer during dinner," but these rules are subject to resistance. The penetration of work uses of information technology into the home leads to an equal penetration at home of the way hierarchies function at work: "I want instant access to you, but I want to minimize your access to me." This strategy increasingly leads to the use of home as an environment in which interruptions can be carefully managed, even among family members and friends. Note the underlying tension in this tale of pressure and resistance in a power struggle between two friends:

I offended [a friend who wanted to] send me a fax. I said, "I don't have a fax." She said, "Why?" I said, "I don't need a fax." . . .The whole thing fell apart because it boiled down to . . . "I am your best friend and you cannot buy yourself a fax?" I said, "Why can't you write a letter? There is mail!" ... Directly or indirectly, it boiled down to [the friend's accusation]: "Why don't you have a cellular?" [and my response]: "I don't want to have a cellular."

Often devices are used to manage relationships by not being used—by turning off the phone, avoiding using cell phones in the car, or checking for e-mail or voice mail only at certain hours. Technical failure can also be brought up to smooth over struggles over access. "My batteries must have been dead" is the new white lie of the information age.

People exchange gifts of information over e-mail and send picture files, "JPEGs" of the baby, and of the house to friends around the globe. Whether in California, Colorado, or Dublin, the pictures maintain a relationship among distant friends. A personal computer given to Grandma ensures that she can receive the JPEGs too. It also guarantees that Grandma will call her daughter and son-in-law if problems occur in downloading the picture files, binding the family even closer together. Cell phones are given to wives to demonstrate solicitous husbandly attitudes—and ensure that they will not be out of touch when needed. Karen and Peter, both engineers, bought her parents an answering machine and cordless phone in response to their own "access frustration." They recall clearly the difficulty they had reaching her parents upon the birth of their baby. But is this a gift for her parents, or for them?

As parents juggle family needs, social divisions are created at work. Parents have one set of problems, courting couples another, while singles bemoan their own distinct woes. Telecommuting parents who work at home yearn for adult conversation. Childless adults cannot relate to the conflicts that working parents experience. A child brought into the workplace when the babysitter is ill and set on Dad's floor with a pile of papers to be stapled, may make Dad's life easier. However, the childless worker down the hall who is trying in vain to concentrate on his writing might not see this on-the-fly parenting as a virtuous act. Jennifer put it this way: "And I personally have felt a little resentment, since I don't have kids. Occasionally . . . if I had to cover for somebody . . . then there's a little bit of resentment." Her sentiments were repeatedly echoed by others.

Parents use technology to mold their children into the people they think they should be. Margaret talks of buying her child " 'Franklin's Reading World'—a piece of software." She adds, "I thought, 'Well, maybe try that.' Occupy his attention. Or I bought him a little plastic laser sword thing. . . . I keep thinking I'm trying to stimulate his mind!" In a social structure in which education and technical skill are the great dividers, gifts of technology are symbols of excellence in parenting. Robert, a civic leader and engineer, talks of the ambitious children he tutors, and encourages parents to buy a computer:

But I think more and more, kids are coming home and saying, "Hey, George down the street was able to whip out his term paper or get access to a chapter on the farm workers' union or Abraham Lincoln a lot faster than I was!" Or, "Hey, Mom, are you going to take me to the library or am I going to get an internet [connection]?" So, I think initially it's the games, but I think the convincing point for the kids is if you really want to do well in school, you either have to take me to the library or get me access to this other thing.

Kate, a high-tech worker in Dublin, repeats these sentiments when she tells of buying her parents a computer and a two-year subscription to the internet. In addition to putting her children in Ireland's "superior" educational system, she inundates them with technology: computers, game-players, and entertainment devices. She notes, "You never burn any of your bridges in relation to technology. Everybody has a skill set, and you never want to lose track of that skill set because you [will need it] someday."

Even a Gameboy is an investment in the future, creating the familiarity with handling technology that presumably gives the technological elite their edge. A study of middle-class African-Americans states that access to the computer gives an advantage, and that "the return on the investment is almost immeasurable" (Muhammad 1996:106). Technological prowess is perceived as the ticket to class mobility.

People use multiple devices and e-mail addresses not only to facilitate being reached, but also to filter out people and organize their social universe. Only a few people may have access to the e-mail address for the inner circle or the family cell phone number. David, an educator, readies himself for the start of the semester, with a profusion of communications and messaging devices around him—pager, cell phone, direct telephone with speaker, voice mail, walkie-talkie, and e-mail. Each device is assigned a specific purpose. He establishes priorities about who has the right of first contact, and manipulates them by using a selected sequence of media. Only support staff use the walkie-talkie; his colleagues leave him messages on the voice mail; and if the phone actually rings, it is likely to be his wife. He manages the input from multiple sources by putting his "on hold" calls on the speaker phone while checking his e-mail to see if there is more information about the crisis mentioned in the voice mail message. A walkie-talkie discussion elicits an immediate response. When his PalmPilot beeps to inform him of an appointment, the communications media fade to the background while he has a face-to-face meeting.

When the people we interviewed imagined future scenarios, they repeatedly anticipated widespread monitoring—using technology to reduce uncertainty. For example, parents dream of monitoring their children's homework from a distance, just as doctors can remotely monitor the progress of a patient's disease. Employers, too, can monitor the performance of their telecommuting workers through the electronic community's infrastructure. Clare, a project manager, told us how she was slowly trying to change the "culture" of the engineers she supervises by importing a software program from her old organization that would make the many milestones—deadlines for particular tasks—transparent to all on the team. Then they could each know how to pace themselves in relation to an external time standard, one that it was her responsibility to set.

This desire to monitor has several dimensions. It implies not only the technological ability to monitor but also access to the requisite sources of data. Such access may be easy to enforce in the workplace, but in the future scenarios monitoring is also extended to the home. We were told that Silicon Valley people must become "masters of the universe," gathering information as needed. A basic issue, then, is the penetration of monitoring into new social domains, with the concomitant assumption that any behavior that can be monitored should be monitored. Just who might perform this monitoring was a subject left undeveloped in the scenarios, but, tacitly, it was plain that parents controlled children, a dominant spouse checked on the other, and middle-aged children could know what was happening to their aging and increasingly dependent parents. These technologically enabled interactions suggest what in China is called *guan*. The Chinese concept of guan means governance, but also suggests benevolence, even warmth. Concepts of control and affection are intertwined. People want to use the technology to enforce control, but they believe their motives to be magnanimous and kind.

Another aspect of monitoring is performance. Those with power monitor; those being observed perform. Knowledge work has inherent ambiguities that monitoring is designed to resolve. As work becomes increasingly cognitive, it is harder to account for one's activities. The work product is not necessarily a material object, and that makes it difficult to measure. While there are periodic performance milestones—tangible markers of productivity—other work practices are more difficult to monitor. Silicon Valley workers talk about the need to create a "performance," often using technological devices as props that will make it clear to the world outside their heads that work is being accomplished. At her high-tech company, Barbara trains incoming admins to "look busy." Organizing, an essential part of knowledge work, is invisible, confined to internal thought processes. Shuffling papers and using devices make that work visible—theater for coworkers to appreciate. In a world of intangibles, knowledge workers need to create a mise en scène by which they can account for themselves.

The dance of control and the appearance of control reflect a deeper strategy in using technology—self-management. Noted as a feature of high-tech work life (Kunda 1992: 57), self-management is accomplished by manipulating technologies to control one's social environment. Access is increased by adding more devices and services. Call forwarding, alphabetic paging, and real time web-based communication allow the world to find you. While an active pager puts a person "on duty" for extended periods, leaving town to go to a concert or take a hike is a form of resistance. People can manipulate the devices themselves to limit access, making sure that the voice mail will store only ten messages and using an automatic responder e-mail—an ebot—to inform the world one is on vacation and cannot be reached. Keith talked about how he once declined a job because it required him to carry a pager, but he will use a cell phone because "you can turn cell phones off!"

People also use devices and material objects to communicate with themselves. Some people leave messages to themselves on their home answering machines or voice mail as reminders. Aaron talks about waking up from dreams in which he has cracked a thorny software problem, calling himself at work and leaving a voice mail. Michelle inserts an exercise period on her "MeetingMaker"—a piece of groupware that allows people to broadcast and coordinate their schedules. She hasn't used that time for exercise in ages, but it holds a place open in her schedule for uninterrupted time when she can actually work. Scheduling devices are also used for reminders to oneself. Control over oneself can be facilitated by the omnipresent devices and artifacts. A beep on the PalmPilot or a MeetingMaker reminder to exercise not only blocks out time but may even nudge the person into exercising. The traditional exercise machine in the family room serves a similar function. Seth, a programmer, decorates his body with multiple piercings to remind himself of his physical body so he won't be adrift in his mental work consciousness. His discomfort becomes part of his self-management.

Trusting a Stranger

Technology enables people to live a "crowded" life, filled with complex and competing obligations. However, where human relationships are concerned, making and maintaining obligations is not a simple mechanical process of "communicating." Trust is essential to making these relationships work (Baba 1999; Castells 1996: 179; English-Lueck and Darrah 1999). Trust relationships are constructed among the members of virtually connected teams by exchanging personal details and joke e-mails to cement ties. Trust occurs when people care about each other and act out of motives other than self-advancement. The reasoning behind trust in Silicon Valley is remarkably coherent. Trust safeguards physical and emotional valuables. Professional reliability is necessary, but it is not sufficient to build trust. Trust is grown by establishing an emotional bond by shared confidences. For network working to be effective, not only must information flow efficiently but, in addition, the people who are communicating must trust each other.

Trust is not an easy proposition in relationships that are largely mediated through technologies across distances (see English-Lueck and Darrah 1999). These relationships are difficult to build using the "chilling" media of e-mail and voice messaging. Communication from a distance makes it hard to know what hidden agendas a person might be harboring. Is the person ambitious or insecure? Is she sensitive to time deadlines or is he an apathetic burn-out who will be difficult to motivate? Is he competent to understand a terse technical message, or does she need to have the context for a cryptic e-mail message explained. Just how much does he know?

Knowing the answers to these questions is difficult enough among coworkers who have daily face-to-face contact. Even there, it is sometimes difficult to cultivate trust. For instance, Stan told us how a counterpart misjudged the problems inherent in changing computer platforms. He described the event to us, highlighting the problems in assuming and communicating technical competence and blame. Several teams had been merged, making it necessary to work across several platforms, PC, Mac, and Unix, to create a CD ROM. Stan's team was experienced in this process and never thought to consider that the other team might not know the tricks of conversion. He reflects on the resultant failure:

And he was sufficiently ignorant of this process, and we didn't think to say, "You do know what you are doing here, don't you?" Because if we had said that, then he would have blown up anyway. . . . So . . . when they didn't work, his first reaction wasn't to say, "What happened?" His first reaction was to say that we didn't test them! Well, we tested them when they were Mac files but [not after they had been converted]. . . . Of course his reaction was to point the finger and accuse and so on. His motive isn't to get the job done, his motive is to not make himself look stupid. . . . That comes back down to trust. . . . I can't trust him to be straight, so you end up with the Machiavellian tricks which I really hate!

Trust, difficult under the best of conditions, becomes even more problematic when professional, organizational, and cultural boundaries must be crossed. Added to this is the difficulty of gathering this social information using "flat" technology such as e-mail, which lacks the capability to transmit subtleties of voice inflection and facial expression. Will some stranger, who nevertheless must be given information, forward confidential e-mail to a competitor? Will these people use information to make themselves look good at your expense?

One way to manage this dilemma is to build trust through face-to-face contact and by exchanging sensitive personal information to forge closer ties. Sometimes organizations design this into a process, bringing team members together for regular face-to-face contact. Individuals might initiate the link themselves over a meal, or a pint of Guinness. Even when mediated over a distance, ties can be nurtured. Reciprocal favors, elicited by phone or e-mail, create a pattern of reliability that underlies trust. Personal information, transmitted over telephone or e-mail, can build on family information shared in person. Sharing an inside joke creates a fleeting bond, which reinforces the budding trust relationship. Trust is built slowly over the course of many interactions.

Trust embodies many work virtues. Trust is created on a foundation of technical competence, a long history of tasks completed correctly and quickly. Most of all, trust provides a cushion of tolerance in a world of distant communication and unpredictable glitches. Instead of presuming a hostile motive when a mistake is made or an effort fails, this web of communication—face-to-face, telephone, and e-mail—grants a measure of forgiveness.

A portion of technological work is always spent struggling, and trusted friends and coworkers can be an aid in that struggle. Telephone systems do not always work. Servers go down. A new, more effective program on the PC means that more work is done on that computer, and so all the data must go off of the Mac and onto the PC, a low-priority task squeezed into a barely manageable schedule. The fax on the other end is out of paper, but an e-mail can't contain the all-important signature. In the meantime, simply learning the various features of a key tool—Microsoft's Excel or Adobe's FrameMaker—may require taking manuals home or struggling through a lunch break to master yet another program's features. Messages are sent to friends in the trust network to elicit suggestions on how to get the new graphics program to work with this particular configuration of hardware.

Unintended Consequences

Engineers have a maxim that each solution generates another problem. The extensive use of technology to solve the vexations of everyday life has generated a new set of unanticipated difficulties. Technology makes possible mobility, which in turn erodes the importance of place. When work can be done in cars, on beaches, and in bathrooms, what is special about workplaces or homes? It confuses the roles that were once associated with those places. The role of the technology, and often of the work associated with that technology, is amplified at the expense of other roles—husband, father, friend. Once the technology has been thoroughly intertwined with family life, it has an impact on the shape of family culture—how family life is lived—and on family identity—how the family, and the people in it, define themselves. Does the creation of life-in-networks dominate social organization at the expense of other forms—family, neighborhood, ethnicity? Working with technology, thinking about technology, and producing technology change the way Silicon Valley people construct reality by giving them new metaphors. What are the consequences of this new framework for constructing a livable community?

Place, and the tasks once associated with place, matter less in the new order. People don't necessarily go to work at a workplace, but carry work around with them as a natural feature of their daily lives. Consider the case of Nicole, a process engineer. She talks about never taking work home, yet her artifacts and relationships belie her statement. The computer has its own room, and engineering magazines litter every flat surface (see Nippert-Eng 1996: 83). Her roommates and fiancé are engineers, and household discussions revolve around daily logistics and shop talk. She separates her ongoing professional education from her immediate job. The former is her own long-term responsibility, the latter is associated with day-to-day company activity. For Nicole, as for many, "work" in Silicon Valley is not a single coherent entity but a collection of different tasks.

People talk of their "work" in a globalized fashion—they have ongoing career preparation, finances, and parenting work. But they distinguish those activities from their "work-work"—that is, paid work for a particular organization. A large proportion of supposedly "free" time is spent thinking about "work-work" while in the shower, eating, or driving. People think about work schedules as they drive to work, check voice mail, or mentally preview the day ahead. They bring the problems of interpersonal office conflicts or critical events of the company into the family dinner conversation. Tom pulls the weeds in the garden, a "Zen-like" distraction, while thinking through a serious operational obstacle at work. Lars reflected on the time his child was born:

The only rule is that I be able to spend time at home, but that doesn't take me away from work, it just takes me away from the office. The week after the baby was born, I was doing work. I wasn't at work, except for a couple of days where I went in for a couple of hours to do something, but I did work. Work is part of my being, so the distinction of place is not as important.

This colonization of life by work is remarked on by many. But it is more subtle than merely answering an e-mail or writing a report at home. Financial management, household maintenance, and continuing education take on aspects of work. Just as a tech writer learns new multimedia specifications in order to develop a CD-ROM manual, parents seek out skills for managing "difficult toddlers." Knowledge workers seek out courses on how to work with people they dislike, or learn Italian to prepare for the next vacation.

In the workplace, relationships are cultivated to create a network of people who can trade technical favors. Similarly, family and friends exchange information and services—she provides the name of a good acupuncturist and receives help in setting up the family's computers into a Local Area Network (LAN). Life is managed and skills are cultivated to fit the needs of a changing daily world.

The logic of work, particularly technical work, involves setting goals and objectives, collecting the skills and services needed to meet the objectives, and producing a concrete "deliverable." This logic can be extended to family and civic life. Kevin, a senior computer scientist, declares that any task that is not spontaneous, but dictates action, is work. To that end, it is work "whether they're aligned to a company, or to a family activity." He adds an example: "I have to set up a party for my sister because it's a planned activity; it has concrete, measurable goals; all of these are all task-oriented things that have certain very similar characteristics. I would call those work."

When "work-work" and life-work are so intertwined, it is even easier to merge the different domains of everyday life. Housekeeping becomes work when the piles on the table contain "work-work" projects to be finished. A draft of a church memo and a shoebox of financial papers sit next to a pile of a tech manager's employee performance reviews. At church Stan meets people who can share their technical expertise, and also help him find good educational material for his children. Is the maintenance of a church network—which includes past, present, and future coworkers—community life or career? As people track and trade their stocks earned two jobs ago, are they doing household financial management, or work? Is that tabletop book on Zen meditations a spiritual guide, or a management tool? Is the maintenance of the altar of artifacts from the American Southwest purely personal, or does it work to amplify a similar shrine in Tom's office that keeps him from feeling bored and overwhelmed at work?

Work and nonwork become enmeshed. Karen, asked by the naive anthropologists if she put her computer to personal uses, responds:

What's personal? That's kind of tongue in cheek, but it's kind of gotten to that. There's very little time anywhere for doing personal things. That's a combination of both of us working and having two kids. So, personal? I'd say the only personal thing I might do is if I'm having lunch at my desk—and I always have lunch at my desk—is I might surf to some site and see what's going on. You know, see what's the latest going on at Microsoft or Sun or one of the other computer companies or competitors, something like that.

Workers are sometimes unable to turn off the emotional and intellectual demands of work. With wry irony, Aaron's work group offers a "Timex" award for the engineer with the most stress-related ailments; it's named after the watch that "takes a licking and keeps on ticking."

People describe coworkers or spouses who use work to hide from social relationships. Leah describes the impact of such work-absorbed men on herself and her daughter:

In my last marriage, I was married to a workaholic. He wasn't able to [put] work aside and come home. . . . So many people are so bogged down with a workload that isn't achievable in a normal number of hours that it's very adversely affecting home life because they spend more time at work, less time with the family, and then when they are home they are so tired from the demands at work that they don't have the energy level for relating. They're exhausted, burned out, impatient. . . . I think especially with my daughter's generation now . . . her marriage failed and a lot of the reason is her husband is a network administrator, and he spends all his time at work. She couldn't get him to come home. . . . Everybody's running around insecure.

People view the devices that import work into the home as central culprits in creating this imbalance. Nicole, about to get married, notes that her intended works at home once a week. She tries to pry him away from the computer. When she calls him, he keeps working, and "it's really annoying to hear, you know, [his] going, 'Yeah,' the clicking, and the 'Uh -huh, uh- huh. What? What? Did you just say something? What was that?'" Multitasking is not such a good solution when viewed from the outside by someone who wants your total attention. Andie comments on her fin de siècle version of the "double standard." Her boyfriend takes work home all the time, but when she brings home some "fun" software to learn, he says, "Come on now, it's my time!" Alexandra, a credit analyst in a large software company, reflects:

There's always a danger of letting work infiltrate in your home life. Bringing too much technology that keeps umbilical cords attached to the office. . . . Uh, just having a cellular phone, or a pager . . . [the cell phone rings and interrupts the interview] . . . You know, a lap-top computer with a radio transceiver, connection to your e-mail, uh, starts . . . making you available to whoever wants to call at any time. . . . I think it really does invade one's home life. And I think home life is for you to recuperate from the many hours, and tensions, and pressures that develop during the day at work, and it's for your children.

However, the technology may be more an enabler than a cause of social fragmentation. Repeatedly, people attributed the fragmentation of family life to the technology. Yet we observed that households using the same devices might use them in strikingly different ways. The same devices have different effects on different types of families. People coming to Silicon Valley from other countries around the world bring their own ideas of the boundaries of family, work, and community. VCRs, karaoke systems, and telecommunications devices can pull together families that are already close—such as the Vietnamese families—while in other families the same devices allow family members to drift farther and farther apart. In one Latino family each new information technology was placed in a carefully orchestrated system of devices that encouraged tightly knit extended family and community interactions. The same devices—camcorders, computers, home entertainment systems—fragmented other families into smaller and smaller interest groups. In Hua's family, an adult son was brought into the parental orbit in order to teach his aging Chinese mother new computer skills. He had the technical expertise to put together a computer system, install the program, and train her in Excel—and thus he was obligated to stay in touch with the older generation. The devices used by each group were the same, but they had different cultural and social expectations about who gets face-to-face attention. These suppositions are shaped by cultural norms—people's perceptions of what "should" be done.

In Silicon Valley, training for technological excess comes early. The technology imbedded in work soon becomes the technology that is part of the family. Erin, the mother of a small child, reflects on how her son reacts to the technology saturation of the Valley's culture:

And then there's the culture of what this Valley is itself, with the use of computers. My son is six years old—we have a computer at home that I brought home from work—and he played with that fairly quickly. He mastered that fairly quickly. We play a lot of computer games, and he does lots of things on the computer. They have a computer at his school, and they do a lot of things with them there. I think that's probably even more so in this Valley. Kids learn some of the computer lingo, and some of the internet lingo, and some of that. The engineers and people that are so entrenched in that bring it home with them and teach it to their children. I think that's something that we are in the midst of.

It is not, however, assumed that adults have all the expertise. Just as the status of young Yir Yiront increased when steel axes replaced stone, it is assumed that the young of Silicon Valley now have an edge. People tell variants of Alan's story: "We had a computer game up for kids. . . . [My friend's] a bright guy, and I'm no slouch. We couldn't figure out how to turn the sound on. [The child] walked up to the keyboard, touched one button, beep, out came the sound. He knew how to run the Macintosh better than we did, and he was only a year and a half. That's scary."

The social order is shaped by technological prowess. Fifty-year-old engineers worry about younger ones who are not weighed down by obligations of family and mortgage, and are ready to replace them at a moment's notice. The forty-year-olds worry about the ones just emerging from a university. High school students look with dismay at younger siblings who grew up thinking of an internet graphical user interface as a natural environment. With pride and worry, elders note that the young own the future.

Family culture and even family identity are shaped by the use of technology. We asked households what it was that made them "a family," more than simply a collection of people. Repeatedly the answer was: "We do things together." To these interviewees, the family is not a natural unit that simply exists, but one defined by common action. Families watch TV, camp, travel, eat, and *talk* together. The devices that facilitate that action or talk—phones, networked computers, pagers, answering machines—take on a serious purpose for these people. Paging your children to let them know you are concerned that they arrived home safely from school demonstrates parental responsibility. Sharing an evening of movies or technology talk provides an opportunity for *doing something together*. Parents sing songs from the Nickelodeon children's cable channel and Disney's film *Mulan* with their children.

While topics of friendly conversation may include music, books, and movies, talk will also revolve around "interesting up-and-coming technologies," and "software found on the web." Karen and Peter, a dual-engineer family, tell the anthropologist about their work spaces at home. They describe sitting side by side on the couch with their laptops: "It's really quaint! . . . Yes, this is how we express our emotives! [They both laugh] It's sick! . . . We talk to our e-mail messages and laugh at them."

Technology provides a mutual framework for discussion, as Priyesh comments:

I find that even with friends, whether they're Indian or non-Indian, it's very hard for me to steer away from a technology topic and talk about something that's actually personal. . . . Let's talk about culture. Let's talk about music. I don't see that conversation going for very long before it returns to something technology related.

Talk about technology provides people with a common ground with otherwise alien adolescent children and elderly but technically inclined parents. Stan proudly notes about his son: "He and I communicate at the level of computers and technology and so on, but also at a normal, human level. He is very good. He is the computer nerd. And he and I do stuff on the computer here and so on." The pervasiveness of technical work creates people who can comfortably discuss technology across generations.

Nathan, an engineer married to a tech writer, reflects on how his familial "culture" revolves around technology:

My dad's retired now, but he worked at [a national physics laboratory] for 170 years [he laughs at this exaggeration], building accelerator power supply magnets and other things. . . . My mom was a programmer before there were programmers; . . . in fact, that's how I got interested in programming. She showed me what she was doing: it was . . . the first implementation of Pascal, and she showed me what it was, and it just made perfect sense. I just read it and I started writing my own programs, and she showed me punch cards and we did some of that and we did some machine time. . . . It's definitely a technically literate family.

Technology becomes the topic of conversation, the basis of jokes, the gift that creates an "umbilical cord" to other people. Joking highlights the significance and contradictions of using technology. People joke about the failures and burdens of technology. Janelle jokes that being on hold on the telephone is the modern equivalent of purgatory. Jokes are regularly exchanged between friendly networkers over e-mail. Prem, a manager in Bangalore, deliberately tries to gather background information on the American political system over the net so he can understand the jokes—all important in establishing trust and rapport.

Septuagenarian Barbara viewed her skill with multiple programs and internet environments as a sign that she was "empowered" and distinct from less technologically sophisticated age-mates. She used her skills to create a new network of technically adept "twelve steppers" who were all engaged in online dialogue, working through the therapeutic procedure established by Alcoholics Anonymous. Her behavior and identity were intertwined with her technology use.

Using technology subtly alters how people experience life. Ruben, a machinist in a high-tech corporation, uses the internet to find inspection criteria, but proudly shows the anthropologist his favorite live-cam web page, a real-time video of Cowell Beach and the Santa Cruz Boardwalk, complete with weather and temperature information. He tries to go to the beach at least once a month, and this reminds him of it. This virtualization of experience is not unusual.

In their future scenarios, evangelists for technology talk about "direct" and "indirect" experience. On the one hand, new technology will permit users a wider variety of experience, but that knowledge will be technologically mediated. It will be controlled by both its producer and consumer, and cleansed of the sloppiness and unpredictability—the "inefficiency"—of real life; it will be "virtualized." The rejection of direct experience was expressed in different areas. Max, a Smart Valley volunteer, thought that virtualization in education would be a boon. He said:

Say you want to become a mechanic. You go to this school. You are working on a computer that has a virtual reality aspect to it and you are looking at a car; . . . via the interactive capabilities, you can test various solutions and actually "do" work on the car via the virtual scenario. What a great mechanic! One who has tested the possible problems . . . as opposed to trial and error. I love that idea!

The saturation of technology in daily life has some interesting implications. As with the presumed masculinity of the Yir Yiront ax, or the value of the imported obsidian tool, the technologies are imbued with meaning that shapes the symbolic world of Silicon Valley denizens. Working in high-tech industries amplifies the importance of technology. Toni talks about her life "swirling" around technology:

Of course I'm selling a technological product, selling digital pictures for God's sake, and I do all my work on a computer, so I need to know as much [as I can] to keep myself going. . . . [My husband's] life is totally around selling and consulting on this automated computer sales process, so basically if he comes home and tells me about his work, it's going to be that some client had a problem with their network and he had to troubleshoot it, and he had to set up their system. . . . Then, plus, [our child] will sit there and play games and then do some online work and, he had a long correspondence with people online, so basically, these computers are such . . . family members at this point. This one [pointing to a new computer] is a pretty new family member, the baby of the family.

Paradoxically, the saturation of technology also leads to its invisibility. People are so immersed in the technology of communication that they forget it is there. Rae illustrates how easy it is to overlook the presence of technology. She wants to actively "resist" its lure. She left a career in high-tech work behind to start a business making custom herbal soap. Yet she blithely talks about how her supplier got her logo off her web page; she does not quite realize that while she may have left the high-tech lifestyle behind, she has not abandoned the technology itself. It is integrated into her daily life as an unquestioned presence.

Technology is not just a set of devices that we use. In a community of technological producers, the very process of designing, crafting, manufacturing, and maintaining technology acts as template and makes technology itself the lens through which the world is seen and defined (see Mitcham 1994: 210). The world "takes on machine-like characteristics" (Markussen 1995: 166) that influence how Silicon Valley people view their own lives. Stan redefines his talent as a systems modeler—the ability to see the world as a three-dimensional process—as the "gift of prophecy" that can illuminate the world for his community. Dan, a retired systems engineer, reframes his understanding of short-term costs and enduring benefits and applies it to social problems. The structure of goal, milestone and evaluation used in high-tech work is superimposed into every aspect of life. Alex, an engineer, says sheepishly:

One of the things my therapist and I worked on was a mission statement—which sounds laughable in Silicon Valley. Dilbert and all that . . . but these were two statements that stuck out from that effort. I enjoy learning new things and teaching others about them, and I would like to try to live a simple and balanced life.

Part of Alex's pursuit of a simple life is using virtual tools, such as online grocery shopping, to make life easier.

Technology permeates the metaphors used by people to describe their lives. One woman, concerned about the added burden of her daughter's athletic practice, worries that it will burn "too many cycles," a phrase used to describe computing efficiency. Efficiency itself is a central value in Silicon Valley life. Harry, a South African computer scientist, quips about having "memory storage problems," and tells us he once asked if some food on a rafting trip "was public domain." The very language used to describe daily life has become enmeshed with tech talk.

The cognitive infiltration of technology goes far deeper than a few amusing metaphors. People transfer the engineering and entrepreneurial approaches to their understanding of the social world. "Instrumentality" and "economic rationality" are philosophical assumptions of engineering that are recast into a notion of social efficiency (see Tiles and Oberdiek 1995: 49). They stop being qualities of technical reasoning and become virtues in and of themselves. "Useful," "efficient," and "good" merge into a single moral concept.

Technology is universally seen as the primary cause of all social changes, for good or ill. This assumption is rarely questioned in our Silicon Valley informants' view of the region's future. Technology defines value-added performance. In the words of a civic worker: "If the technology falls through then we just become a bunch of blue-collar workers, not so innovative. I don't know how to describe it, just, well, more like the rest of the world." In the best future scenario, technology would relieve overcrowded schools, erase cultural differences in the classroom, and even relieve the managerial class from the necessity of experiencing other cultures by replacing disorienting and time-consuming travel with efficient telecommunications. Technology would bind fragmented families together with pagers and cell phones that would allow mobile family members to remain connected.

Technologies such as the PalmPilot, the embodiment of the dream of organization, are viewed with admiration, as in Jeff's comments:

This is the first PDA [personal digital assistant] I've ever had, and I delayed a long time in buying anything of the sort. This was only four hundred bucks, which is much cheaper than a Newton; I had to think a long time. I can afford a certain number of toys, but four hundred bucks is a lot for a toy. And I had to convince myself, and to a lesser degree [my wife], that this was something that would be useful to me and wasn't just a case of techno-lust. . . . I finally decided a little fearfully that it's something that would do me good, and I went out and I bought it. And, man, has it done me good. . . . After watching me use it for eight months, [my wife] has bought one and it has instantly become indispensable to her. . . . But we both found it tremendously useful. We're now drooling over the new Pilot which has the infrared beaming ability which will hopefully allow us to coordinate our two collections of things to do much more easily. At this point in our life we've acquired a habit of every couple days of sitting down making sure everyone's calendar and shopping list are up to date. . . . The computer is a tool for the mind. And this is really true. You know, once in a while someone who isn't as sold on all of this stuff as I am will say, "Why do I need a computer?" And the answer is, "You don't need a computer." Nobody needs computers. But if you've got a computer, there are so many wonderful things you can do with it to either streamline your life or to do things you couldn't do before. . . . This all helps save us time and get things done more efficiently. . . . If we're saving all this time, how come we don't get enough sleep?

It is the promise embodied in the technology, not the delivered efficiency, that makes the information tools so culturally powerful. Jan, who worked for the Smart Valley initiative to encourage local internet use, nonetheless cautioned against mistaking speed for virtue. He notes that efficiency may mean "driving two hundred miles an hour on the wrong road." However, the corollary virtue of effectiveness is still one deeply ingrained in technological process. Planning, self-direction, and technical expertise are deeply valued. These values are not unique to the culture of Silicon Valley, but they do embody part of the intangible emotional tool-kit that is associated with technological saturation.

Valuing efficiency, working for the future, and saturating daily life with technology are diagnostic features of a self-identified Silicon Valley way of life. These values play out in interesting ways. Even those people not directly working in high-technology industries become associated with it. Local anthropologists reinvent themselves as "anthropologists of technology." A psychiatrist, one of Sandra's neighbors, talks about himself as "in high tech" since "those are the only patients he gets." Administrative assistants, machinists, and gym owners all view themselves as part of the "high-tech culture" by virtue of their support of it, or through their customers' connections. Chris, a research scientist who works with Europeans who have come to consume Silicon Valley cultures, notes:

I don't have any real strong friends who are "sheer fuzzies." . . . I don't think they really exist in the Valley. They can't afford to and they don't really enjoy it out here. . . . The people up there are still more technical than you'd ever find in the real world even when they are artists or they're writers. . . . They know all this kind of stuff so you can always talk, because computers have become main-stream. That used to be geeking whereas now it's very normal. . . . They're not technical in the common sense of where they're educated, but they're techies of some sort.

While this is an overstatement, many people are touched by the culture being shaped around them. If nothing else, the high cost of living and the low employment rate experienced in the late 1990s and early twenty-first century have a tangible effect.

The less tangible values and mindsets create a technological problem-solving approach. In other social settings, different approaches are emphasized. Devout Christian fundamentalists frame the happenings of their world as functions of "good" and "evil," enacting a "religious problem-solving approach." During China's Cultural Revolution, every action was viewed as a political event—even choosing a bride, who must come from the peasantry or proletariat class, and be politically pure. In Silicon Valley, people view the daily conflicts of life as "social engineering problems" that can be "solved" if given a thoughtful and systematic appraisal.

In the next chapter we will explore the many ways in which Silicon Valley people mold their relationships into effective instruments. We also will see how the technological problem-solving perspective leads

residents to "engineer" a value-added community. One of the key implications is that the suffusion of technology creates an alternate framework
for identity, and affinity. It forms the basis of an ethos of instrumentality
that permeates social relations. Building and maintaining networks becomes an integral part of community.

Networking

Building Community in Silicon Valley

Doing Lunch

Sheila has to get ready for her lunch gathering of Las Madres. Five moms are coming to her house today, all part of the neighborhood playgroup. Her little group is part of a network of more than thirteen hundred mothers in Silicon Valley (San Jose Mercury News 1998b: 13E). Sheila has to get the house picked up and make some light lunch snacks acceptable to small children and their more discerning mothers. Bagels will do for the former. Peets coffee, Greek salad, and homemade cheesecake will satisfy the moms. All of the women have children born in 1996. Sheila has two kids. Her daughter is a preschooler and her son is much older, already entering sixth grade. Also coming are Joanna, Diana, Rebecca, and Melina. Not surprisingly, all have multiple associations with each other, including connections to the high-tech world. Joanna is now staying home with her young children, but she wants to return to tech writing, or at least writing, professionally. Diana is likewise at home now, but plans to return to software engineering. Rebecca juggles motherhood while working flex time as a software engineer. Rebecca, Sheila, and Diana often break into shop talk, particularly since Rebecca can keep the at-home moms abreast of what's happening in the industry. Rebecca also is in Sheila's investment club, and Diana goes to her gym. These are typical of Silicon Valley interconnections. Sheila and Rebecca also belong to a reading group that meets once every two months to discuss a book of fiction. Sheila likes science fiction, as does her husband, Eric, but being in the book club forces her to broaden her interests and meet different kinds of people. Her group will be reading a Vikram Seth novel next. She is keen to read it, since it will help her understand the Indian culture, which has a strong presence in her workplace. Melina, a researcher at an R&D firm, is very pregnant with another child and concerned about the stress of juggling yet more children with her work.

The Las Madres group is having a Halloween get-together next month, and that means getting the old sewing machine out and making costumes. Rebecca silently wishes she could just buy a costume, but she wants her child to be able to remember someday that even though she worked, she took time to make one by hand. Joanna has already made a cute chili pepper costume. Maybe Sheila will sew a costume for her daughter to make her a complementary condiment. Sheila's son will undoubtedly have his own plans for Halloween with his friends. His friends have also talked about having a Star Wars group, so maybe she should work on a Darth Maul outfit. He would find that "cool."

As they are preparing their children's bagel snacks, Rebecca and Sheila talk about the politics at Rebecca's workplace. Rebecca is working on a C compiler for her Cupertino company. Diana notes that Rebecca's project has just been derailed. Somebody "goofed up so she had to put in extra hours to stick to the schedule," which explains why Rebecca has been so scarce lately. Maybe in a year or two, Sheila will go back into industry. She just needs to stay on top of the technology—and the corporate politics.

The kids go out to the play equipment in the backyard and the moms get down to some serious noshing. They talk about their parents, their children, the new park in Campbell, and the possibility of making a joint skiing trip this year to Tahoe. Rebecca has a timeshare in a cabin there. Melina, pregnant now, thinks she won't be skiing this year. She tells everybody in the group about the newest piece of equipment at her workplace, a fancy breast pump. Fully one-half of the women in her company, from admins to managers, are either pregnant or have just had a baby. She overheard her colleague in the next office joking about putting a contraceptive in the water dispenser. While she is willing to use the new breast pump, Melina notes that her coworkers give less-than-satisfactory reports about going to the women's bathroom and pumping out milk to send home in bottles to the various nannies. It is an ordeal, but at least the company thought enough of them to provide a chair and the pump. The talk turns now to the kids.

Sheila's daughter Jessica goes to a co-op preschool, so Sheila has to contribute her share of time as a coteacher. She is responsible for all the craft preparations—designing sewing and fabric projects, gathering materials, and typing up directions. Appropriately, before the birth of her second child, she used to work as a process engineer, and this is not too different from that work. Two fathers work with her at the preschool. One does the techie maintenance and the other is her "handy" counterpart, a project designer who develops carpentry and other tool-based projects for the kids.

Diana is in a similar situation, acting as cooking director for her child's preschool. She is the repository of information for her Las Madres group. She reads books and finds new web sites on child rearing and shares what she reads with the group. She is concerned about the latest information on pediatric health care, discipline, and early childhood education. Diana applies herself to the task wholeheartedly, knowing that for now, it is her work. Even as she surfs the web each night, getting parenting information and some recipes for the preschool, she is keeping her hand in by visiting the technical sites that might help her get her next work. It might be difficult to get back in after having taken a few years off, but there is always room for competent engineers—at least in boom times. Today her life is taken up with the daily details of parenting, but it doesn't hurt to keep her eye on the long view.

While the kids play happily on the deck, Sheila goes to her husband's computer to show the other mothers a new web site she has found, using her husband's computer, on things to do with children in the Bay Area. The computer is the latest and best Apple has to offer. Beside it is a box full of envelopes and papers—mostly bills. Things are a bit tight for Sheila and Eric now, even though he has a lucrative job as a project manager. His skills as a software engineer are excellent, and he is always anticipating the next direction in programming and learning new things. He is a dedicated father, reading to his children each night, using his flex time to coach his son's soccer team, and devoting much of his weekends to his kids—at least while they are awake. Attached to the monitor are pictures of their two children at various ages, the older child in soccer uniform, the younger daughter dressed up as a fairy.

Sheila remarks that they worry a little about her son, Brian. His grades are good, especially in math. Brian would live on the Sega Play Station if he could, but both parents insist that at least part of the time spent on games be spent outdoors or on the "real" computer. Sheila is proud of her husband. Eric not only does his work well but also participates in his company's culture. He arranges events with his staff, such as picnics and outings. His team has been working hard, so last Friday he took them all to a matinee. He volunteers in the corporate philanthropy program. As long as Sheila is staying home, they have to watch their cash—mortgage, college funds, and the like take their toll—but he can give his time. The company has a program that matches community needs with available staff skills. He has worked for schools, the Tech Museum of Innovation, and various charity functions, writing pieces of software to make their work a bit easier.

On the computer desk is a cup full of odds and ends, including a "NetDay" button. Brian's school had a huge NetDay III event a couple of years ago. Eric went into his son's school, along with about eight other parents, and helped install donated computers, hooking them up to the internet and making sure that the new operating system didn't eat the old programs. For his efforts he got a nifty T-shirt from Madge Networks. The year before his NetDay, Vice President Gore came to participate in a much more publicized NetDay event on the Peninsula.

Eric makes sure his kids have every advantage. Jessica has had her own webpage since birth, and Brian cannot even imagine doing homework without consulting the web, not to mention the time he spends on the various Lucasfilm web sites. But not every kid has those advantages. Eric has been telling Sheila he may spend even more time on the philanthropy council at work. The other women nod; their husbands are like Eric in this regard. Eric has made some good connections through his volunteer activities and feels he is making an impact on the community. He likes the idea of Joint Venture: Silicon Valley fostering efforts such as Smart Schools and Smart Valley, making a difference in the way the community is run. His own company has given more than a million dollars to help link public school students to the internet. It doesn't hurt that those kids will then want to get computers connected at home, making this place truly networked. One might as well think about the future and work on it now.

The women mill around the den, where they have been eating and watching the kids. On the wall behind Eric's work station, Rebecca notices a scrap of paper attached to the cork board. Eric has printed out a quotation from his hero, computer science giant Dick Hamming of Monterey's Naval Postgraduate School. The saying reads: "If you don't design from the top down you will hate yourself in the morning."

Identified by Technology

The notion that a culture can be identified with its economic specialization—and the technology associated with it—is a very old and widespread idea. The Chumash, the native people of the Santa Barbara region
of California, were so named because they were the "people who made
shell money," the strings of beads made from olivella shells (from the
marine purple olive snails) that were used as currency throughout the region (Miller 1988: 111–112). The Chumash carved a niche out of the regional economy, creating a medium for exchange that allowed people
from the Channel Islands to exchange resources with coastal mountain
neighbors. Whether it is mining coal in Newcastle, processing cattle in
Kansas City, or making wine in the Champagne valley, economic specialization has often provided an identity for various communities. Mention Motown or Motor City and Detroit immediately springs to mind.
Hollywood and Hong Kong conjure visions of filmmaking.

In all these places, while no one would postulate that everyone in the
community directly participates in the key economic activity, it is that
activity that shapes the community. Among the Chumash not all were
merchant money makers, but the value of money was set by the *'antap*
shamans who also figured in the choice of political leaders and were key
players in the *tomol* boat societies that composed the Chumash elite. The
making of shell money established a whole way of life that was developed around the exchange of resources, supporting the creation of a social elite that was in turn supported by a whole worldview. Trade and the
religious elite justified the social order. There are people in Detroit who
do not participate in car production, but the peculiarities of the automo-
bile industry still influence the lives of waitresses and schoolteachers.

The high-tech industrial base in Silicon Valley is based on such industries as research, development, and production in aerospace and defense;
semiconductors; computers; software; telecommunications; and biotechnology. All of these have played a part in the evolution of its economy. In technologically saturated communities, the daily pervasive economic presence of technology defines people's social relations and images of the world. On the individual level, people extend the logic and
methods of project-based work to the rest of life. We see this process ex-
emplified in the actions of Jonah, a software engineering executive. He
was compelled to travel to Italy to complete contractual obligations. Ini-

tially Jonah was determined to be miserable—prepared to loathe the experience and hate the food. Then he turned his ability to manage projects to his own problem. Jonah changed his "set of goals," borrowing from the problem-solving approaches he used as an engineer in high-tech work. He systematically set out to try different kinds of food, to learn Italian, and to expand his horizons. He redesigned himself to fit the new specifications.

At a larger community level, people who live in the region also use the models of high-tech work to formulate their understanding of community. For example, media and civic leaders talk of Silicon Valley as if it were a giant symbolic operating system, connected to diverse, distinct cultural applications programs that function on its platform. Just as Windows, or Linux, allows people to run word processing or spreadsheet applications, an educated work force with the appropriate work ethic would "provide a platform" for newcomers, whether from Vietnamese or Midwestern cultures, to take root and flourish. A technologically saturated community develops a distinct ethic and civic discourse. People's thinking about their own life trajectories and community values are filtered through lenses shaped by technological metaphors and problem-solving perspectives.

A technologically saturated community also has a distinct social profile: from labor relations to family roles, the social order is realigned to serve the industry. The only "natural" resource that Silicon Valley clearly identifies is its expertise. Only that density of knowledge will maintain Silicon Valley's leading role in the volatile world of technological capitalism. Education, particularly the kind of education that provides vocational and technical skills, will produce engineers, teachers, and burger flippers with the drive and capability to create the "value-added" that distinguishes Silicon Valley from Lincoln, Nebraska. Individuals, organizations, and communities can increase the worth of products by virtue of their expert involvement with the production. Silicon Valley workers believe that their distinction is based on technology, or more precisely, on a worldview in which technological metaphors and models predominate.

The core of high-tech workplaces has a ripple effect, reaching into the lives of people who are not directly connected. Nontechnical workers must adjust to the lives and tastes of their affluent "techie" clients. Gym attendants and massage therapists struggle to keep the stressed high-tech workers who are their clients healthy and fit for creative work. They must adjust to the clients' sense of tightly compressed time. More important, they must share the clients' belief that the self is a means of production, that maximizing the productivity of one's body and mind is yet another project. Artists, too, must cater to the expectations of their high-tech patrons who adorn corporate walls with their works. Living in a region suffused with technology, it is no surprise that the artists incorporate technological media, weaving together traditional and computerized techniques.

Even the most traditionally menial tasks are seen to be completely transformed by technology. Angelina, a director of the Center for Employment Training who trains the next generation of janitors, made this observation:

In terms of the custodial, you're looking at some incredible equipment that is now available. It's not the mop and broom anymore, of course, and it's not just the simple buffer. This is equipment that is really, really specialized, that also can be computerized, and we're looking at hopefully writing some grants because we are nonprofit, to make that equipment accessible to those custodial students. So it's not, you know, a simple class anymore. You're talking about chemical usage and the mixing of chemicals and knowing exactly what each of those chemicals is. And I think you're going to see, because of consumer awareness and environmental protection, you're going to see changes in the future in terms of cleaning solvents and solutions, and that we also want to prepare students for as well.

Angelina's belief that life, even the life of the humble, is being transformed by technology is a hallmark of Silicon Valley's technological saturation. Even those seemingly disconnected from the glamour of high-tech work have a heightened awareness of devices.

Families are reshaped by the technical workers in their midst. When a spouse travels to India, or a father must work a hundred-hour week, the lives of nontechnical family members are affected. Hua, a Chinese immigrant who had worked in garment industry shops, now works on a high-tech production floor. She imagines a more genteel work life, because her son, reared in the technologically rich education system of the Valley, can teach her to be computer-literate when he comes to her house after work.

Civic life also illustrates the alteration of nontechnical realms by high-tech work. Nonprofits network talented high-tech volunteers with worthy community programs. The functional intertwining of municipalities into the region—each with its own environmental and building codes—made it difficult for a business to have one building in Cupertino and another in Fremont. High-tech businesses urged the governments to cooperate to create a common set of regulations, available online, making it easier for businesses to expand in the region. Municipal governments then streamlined regulations across the many jurisdictions in the Silicon Valley region to promote regulatory "one-stop shopping."

The towns and cities at the edge of Santa Clara County were the original geographical area identified with Silicon Valley. As the economy expands, Fremont, in neighboring Alameda County, aggressively adds itself to the list, hoping to attract new business. Gilroy, still known for agriculture and garlic production, has become a bedroom community for high-tech commuters. Even San Francisco, initially reluctant to identify itself with the Silicon Valley phenomenon, admits to its connection through its many Multimedia Gulch internet businesses.

The passion for the production of high-technology commodities shapes the public landscape. Everyone in California has seen billboards trumpeting "Got Milk?" or "It's the Cheese." But only in Silicon Valley can one see the cryptic billboards promoting "FPGA 2 ASIC" (Field Programmable Gate Array to Application Specific Integrated Circuit) or Java computer language. In the minutes before the beginning of the feature film, movie audiences view advertisements luring technical talent to particular companies, and people joke about exchanging business cards up and down the theater rows.

The regional social order reflects the dominance of high-technology industries. Service providers—educators, restaurateurs, and janitors—are there to serve the needs of the techno elite who really "matter" in the Valley. Any amenity that enhances quality of life—clean air, good educations, reasonable transportation—becomes part of the "value-added" that can be cited by companies to recruit and retain the expert workers that are valued as the true Silicon Valley natural resource.

Designing New Platforms

Life is not just lived in Silicon Valley; it is, at least in part, designed. While people experience their daily lives at the individual level, as in the stories you have already read, there are social institutions that operate beyond the individual. Technology saturation of Silicon Valley infiltrates personal choices and actions, but it also acts at a broader level. How does the culture of Silicon Valley affect social entities beyond the individual level? How do networks, work organizations, and civic organizations support and shape the region? The civic fabric of the community at large reflects conscious group efforts, not only the sum of random individual actions. The area is not a mere collection of individuals with their established separate identities but is a composite of multiple shifting communities, networks, and alliances. New identities are forged around the compound identities and distinctive work that brought people to the region. Symbols are used by individuals to shape meaning, but they are also used by groups to define the whole region. Technology thus has implications not only for individuals but also for the whole community, as Silicon Valley people learn to live in a culture of their own creation.

The process of creating new culture is called ethnogenesis. Innovative economic patterns, technologies, and social organizations are invented to replace or augment the old ways. A distinctive ethos comes first to be associated with that way of living, and then to actually define the new society. As I cast about for cross-cultural examples, I found historical frontiers to be rife with instances of ethnogenesis.

When a population moves into a new territory, it lacks the resources and often the desire to duplicate the parent society. Economic specializations are impossible to duplicate in the new sparse setting, or are wildly inappropriate. The old social order is insufficiently reproduced, and new relationships and values are instituted. There are many historical examples of this process. In the post-Roman frontier society that divided early medieval Europe, extending the social order of "civilized" Europe was no longer simply a military exercise to extend the political economy of Rome. The infrastructure for such a wholesale conversion was lacking. Instead the agents of cultural conversion were linked to the Holy Roman Church.

In the fifth, sixth, and seventh centuries monasticism provided the incubator for a number of cultural inventions that transformed Western Europe. The "outposters" of monastic life did not simply duplicate the social order of the societies from which they were drawn; they created new structures. The monastics established the Benedictine rule to create islands of industry and cultural reproduction. The new monastic social orders marked "milestones in economic history," radically reorganizing the patterns of work practices (Sullivan 1979: 33). The new orders mandated manual labor for people from all classes—even the nobles—some of whom would never have known such labor before. They also introduced a civilian lifestyle regulated by schedules, heretofore a military concept.

The religious orders invented new managerial techniques to govern the monks. Hagiographies, the stories of saints' lives, created a lore of leadership stories that defined what it meant to be "good." The ideals of denial of "human nature" and submission to God reshaped the qualities expected not only of holy men but also of kings, warriors, and workers. The medieval ethos, a pessimism that mistrusted human frailty and advocated utter dependence on God, flowed from monastic ideology. That ideology underlay a religious epistemology, an explanation of the universe, and a religious problem-solving perspective that dominated both medieval intellectual endeavor and the actions of people in everyday life. God, not the chemical properties of willow bark, relieved pain and healed the sick. God dominated discourse, from political pronouncements to the profanities that invoked God's body parts. Monks were "Christ's athletes," an image that suggested action and sanctity as models for community. Born of frontier innovations, the social organizations and beliefs of the monasteries thus transformed historic Europe.

In Silicon Valley, ethnogenesis takes a different form. Corporate entities and transorganizational networks dominate the social order. The heroes are technological innovators, whose admirable qualities are creativity, efficiency, and entrepreneurship. Elliot's account of Silicon Valley's success reflects this admiration:

We're optimistic here in Silicon Valley because we have no reason not to be. I mean a little company called Fairchild Semiconductor started messing around with these integrated circuits, and boom, Silicon Valley happened! And then some researcher in Stanford started messing around with the DNA molecule and splicing it, and boom, we have the biotechnology revolution! . . . The ground zero is right around Stanford, the epicenter, and . . . the boom can be heard around the world. So sure, we're optimistic about technology. The internet . . . that basically started here with Marc Andreessen [creator of the Mosaic browser and cofounder of Netscape], and what he did for providing the graphic user interface. I mean the ARPANET had been around for decades. But who used it? You know, a couple of researchers? Then you provide a graphical user interface, and boom, [he laughs] the world is using it!

The accounts of heroes foster an optimistic worldview, dramatic and successful. The engineering skills that underpin that heroism are dominated by planning, design, and other future-oriented activities. These qualities are imbued in the models of leadership that extend beyond industry into education, governance, and even religion. They create a distinct problem-solving perspective in which innovation is favored over following the tried-and-true. Technological solutions are promoted for a variety of social ills, from poor health to illiteracy. Silicon Valley is particularly vulnerable to the " 'dazzle' effect—presuming that the best solutions are the most technological" (Hakken 1993: 118). Technology is "cool," and to be a part of it, even tangentially, is "exciting." Engineering thinking processes link function with structure, what things do with what they are. This framework is applied in turn to human relationships, instrumentally linking what people do and who they are, making efficiency, networking, and innovation inherently admirable.

The chief agent for the penetration of these notions into the culture is the workplace, Silicon Valley's counterpart to a medieval monastery. Silicon Valley has an array of corporate cultures—each with its own mission, structure, and "culture of values"—that are shared, if only temporarily, by its regular workforce. Temporary workers may partake of the culture, hoping to use their participation in it as a vehicle for promotion. Immersion in the corporate culture, not always possible for telecommuters or outside contractors, is viewed as the instrument for enculturation and success. Some companies, like Apple, view their culture as a key commodity. While the new monochrome logo makes it no longer necessary to "bleed six colors"—a term borrowed from the old rainbow-colored logo that was once used to describe corporate loyalty—work at Apple is viewed in missionary hues. Beth, a researcher, comments:

Being in Silicon Valley, it's part of a culture of people who put their heart and soul into their jobs. . . . [It] seems to be more socially conscious. . . . [Y]ou think about how the place you work affects the community or affects the world. . . . When I first [worked at] Apple, we felt we were changing the world. At Apple you definitely have the feeling that you impact people's lives.

David Packard's book, *The H-P Way* (1995), recites the history of two men working out of a garage as the origin-myth of a philosophy of democratic and inclusive management that continues to guide the organization. Such companies define a work style, a mission, and an ethos that tries to transcend mere employment, taking it to the level of social transformation. As I noted in the previous chapter, people are changed through the process of management training sessions and by the process of management itself—setting goals, meeting objectives, and anticipating market changes.

Jerome, a veteran electrical engineer turned executive, predicts that work will increasingly demand that individuals manage themselves or risk being completely consumed:

We'll get up, shower, brush our teeth, have breakfast [and start working]. There would be a lot of information sources available. There is a very . . . compelling scenario that work could become a twenty-four-[hour] experience, which, I think, would be horrible. But the information technology will allow you to do that. So, I think it will be dependent on you and your own self-discipline to separate the church and the state. [Each one of us will] have to separate what I'm going to spend on work and what I'm going to spend on my personal life because lines will blur.

Workers in Silicon Valley are viewed as mobile "bundles of skills," and they must constantly work on themselves to meet changing demands. This "work" includes self-marketing, keeping one's résumé updated and one's social network intact, and accurately anticipating the changes in future products and the skills they will require. Jeff, a survivor of many project closures, notes:

So, you know, the companies are always looking for the next product or the right technology or the right way to market what they've got and to fight down their competitors and keep their niche. And their employees are doing exactly the same thing from a different angle. They're trying to find the right job within the company to keep their job security and to keep their career nicely burnished and looking good. You make a wrong decision, a wrong prediction, about the market, and suddenly you're in a backwater product—[the] project that gets canceled and becomes unimportant. And the skills you spent two years investing in it are not interesting anymore. A bad guess can leave you kind of stranded, and it can be a real hunt to find a project that will . . . energize your career again. Which way am I gonna jump? What's the market going to look like in two years? Is this the hot project, or this going to be the one they suddenly realize was a total waste of fifteen million dollars and gets canceled next summer?

The prediction of the future job market and upgrading of skills become the leading passion of workers in the Valley. It also makes it clear that the burden for maintaining the social contract between worker and company is placed on the individual worker.

The ambivalence inherent in the relationship between worker and company is seen vividly in their discourse on loyalty. Job mobility drives a social contract that is increasingly one of mutual instrumentality. Turnover in Silicon Valley is twice that of the national average (Joint Venture: Silicon Valley Network 1999: 10). Loyalty is not viewed as a lifelong contract between workers and companies, but a temporary arrangement. Workers have learned the hard way about the realities of the social contract. Jeff again notes wisely:

Being laid off feels like being back-stabbed, but you have to keep the company afloat, so—I tend to be forgiving about that. There's a little cynicism worked in there. You go to work for a company and they come at you all "rah-rah" and they talk about "team spirit" and the "company loyalty." . . . There is such a thing as company loyalty, and it can work both ways. But you have to remember that you're working for a business that's responsible to its shareholders. And loyalty stops there. If they stop the money, loyalty goes. And that's . . . fair. If [my old company] was loyal to all the people it had laid off and not laid them off, it would probably be gone now. And we'd still be out of a job. That's tough; that's the realities of life. [The company] was loyal in a different way and gave us nice severance packages. I worked for a company once that tried intensely hard to be loyal to its people and did go out of business.

The need for workers, who may return as contract workers after being laid off, has led to a climate in which layoffs must be handled gingerly. "Most companies are now trying to keep their cool in the exit process. Ex-employees don't want to burn any bridges either. And employee recycling has become an ecological business necessity" (Ewell 1997: 1A, 8A).

Multiple Linkages

Because the relationship between worker and company is ephemeral, an alternative organizing principle must be created. Instead of building relationships around workplace organizations, people build their relationships around work networks; that is, they are centered around co-workers rather than companies. Anthony, an information systems applications architect, has found that "the bonding of people into groups, particularly groups that work together . . . survives within a company and across companies and that those bindings are much stronger than the company. And that because people moved around so much, the bonding now tends to be amongst people, more than it is among companies."

Jill's story illustrates the power of networks. She is, or rather was, a marketing director for a large high-tech company. After she originally started with the company, members of her network cohort were then hired, one by one. Within her unit, from admin to VP, the team was assembled from nodes in her personal and professional network. When the company reorganized, it reversed the process, and one by one the members of the network were laid off. In the end, only the admin remained behind from the original group.

Michelle illustrates another kind of network. Years ago her friends were all employees, or spouses of employees, at a large software firm. Most of the women and their husbands left that company, the majority going to another firm to form its upper management. The women still meet for "girls' night out," celebrating birthdays every month or two. Their conversation about their diverse companies' business, children's achievements, and interpersonal relationships reinforces the network. They help each other into jobs and through career decisions. Michelle also creates other networks that serve her needs as a working parent. At work she hired a contractor who, like herself, was the parent of two small children. That connection allowed her to locate a desirable pre-school that covered the 8:30 to 5:30 time slot, had educational activities instead of television, and was conveniently located on her way to work. Given the difficulty of locating suitable day care, the information Michelle received about the center, and her introduction into its network of parents, was worth as much as a tip leading to a new job!

There are many different kinds of networks. A network is not a simple homogenous entity. Networks can be informal or formal, short-lived or enduring, distant or intimate, or all of the above at different times. Networks formed of company "vultures" who e-mail each other whenever food is left over from anyone's team meeting are ad hoc and ephemeral. The coworkers that function as fictive kin at holiday celebrations, joining together several nuclear families in decades-long relationships, are a different kind of network altogether. The quality common to networks is that they are both useful and supportive. The network is a form of social organization that marries sentimentality and instrumentality. People care about the other folks in their network, but that does not keep them from finding opportunities for utility.

Networks can be consciously used to facilitate work. Consider the following example. In a phenomenon created by the region's exorbitant housing prices, a group of young professionals—four men and one woman—are living in a group home. Three of the people work in start-ups, Sidney works at a research corporation, and one of the people works at Netscape. Sidney explains how one of his roommates, an entrepreneur, started out at Apple. He and a group of fellow employees split away to form a start-up, each one maintaining his or her connections back to Apple. Within six months all those people had gone to other places to work, so they extended their professional connections to those other organizations. Six months later they had gone to yet other places, extending their work connections to yet more organizations. All of these relationships could be exploited for their joint endeavors, so without having to actively pursue more than their internal connections, they managed to sustain a business network.

Silicon Valley folk live in a complex web of strong ties, formed by intimate friends and family, and weaker ties fostered by organizational affiliation. Communities of practice—people bound by common activities within the worlds of work, worship, and recreation—may yield close ties or distant ones. Some relationships are clearly instrumental. Sylvia, a tech writer, cites her career coach and her therapist as her most significant relationships. These relationships have been carefully cultivated through years of attention.

One of the startling and significant features of Silicon Valley networking is the ease with which people create casual relationships, exchange useful and pragmatic information, and then part. Consider Carl's comments. He works in a startup, gathering information. Much of his work is done on the internet, but he needs a richer environment in which to incubate his ideas. Carl says:

I can walk down to any pub, any deli in town, and sit down and hear conversations on the latest technology. I can bounce ideas off other people. More than a couple times, I've gone down here to one of the little restaurants down the block, sat down and started talking with the guys at the next table, and beating things back and forth. They're with a start-up company, and I'm with a start-up company, and, jeez, there's some things we can do together, there's some ideas that they tried that didn't work, or various other things.

Other relationships are more organic, but still useful—Barbara helps her daughter get a job at the company, Tom's wife gives sound advice on managing problematic relationships at work. The two approaches to instrumental networking coexist. Some cultivate strategic relationships and others use networks "inherited" from family, friends, and classmates. In light of this web of individual relationships, the work organization is almost incidental, becoming just one venue among many for networking. In the words of John, an entrepreneur in Dublin, Ireland, networking is essential to establishing an identity in Silicon Valley. He calls it "the Silicon Valley two-step. Just dance with everybody and eventually you are a player."

Networks can be more formal. Organizations provide support groups for entrepreneurs and business people. The Small Business Institute offers a twenty-six-week program to low-income and dislocated workers to train them to become entrepreneurs. Other organizations, such as the United Minority Business Entrepreneurs, help women and minority business owners win governmental contracts. The San Jose Downtown Association is rooted in place; its aim is to foster business growth in a specific area. Financiers from the Venture Corridor along Menlo Park's Sandhill Road—the source of 30 percent of America's independent venture capital (Wasserman 1996: 21A)—meet at the Quadrus Club to eat and scribble ideas on the paper-covered tables. Retired executive associations, the YWCA, and local colleges all provide further venues for network-building. As we will see in the next chapter, ethnic organizations provide a basis for networks for particular groups. Within companies, ethnic groups, speakers' clubs, and athletic facilities provide yet another opportunity for networking.

Architects for the Company Town

At the community level, workplace organizations have a tremendous impact. The restructuring of community discourse and action around business is one of the major innovations in the social organization of the Valley. Community partnerships forge links among nonprofit and governmental organizations and the financial, political, and knowledge capital brokers of industry. The Joint Venture: Silicon Valley Network is the most prominent of such partnerships.

Joint Venture enlists its members from municipal governments, high-tech corporations, the media, nonprofit and educational institutions, and a variety of business-oriented networks. When recession hit in the early 1990s, the organization responded by proposing that the region boldly "reinvent" itself (see Saxenian 1994: 163–64). Using the language of engineering, entrepreneurship, and design, community issues such as housing, transportation, education, and recreation were recast as "value-added" factors to be used to recruit new businesses and workers. A series of initiatives was developed and enacted to improve features useful to the residential workforce—preferably by adding more technology. In such partnerships networking is leveraged by the wealth of corporations to make manifest a vision of the future (see Glatstein 1994; Joint Venture: Silicon Valley Network 1993, 2000; PRx, Inc. 1993; and SRI 1992).

The visions articulated by these organizations reveal the assumptions behind the design of the Valley. In the earliest iteration, which coincided with the economic downturn of the early 1990s, the scenarios given by the organization closely correspond to the best, worst, and most probable visions of the future expressed by our informants. The scenarios were designed to mobilize a complacent community to action. The most optimistic vision, given a "win-win" score by the organizations, is the "American Technopolis," in which regional innovation is sustained and secures a strong economy. In this scenario, the local workforce is comfortable, secure, and better equipped to serve economic growth. In short, the company town is perfected. The worst case formulation is "High-Tech Manhattan," in which high corporate profitability, without a technological edge, results in a declining economy and collapsing infrastructure. The future seen as most probable is the "Virtual Valley," in which decentralized, "lean" enterprises change swiftly to service new niches. Growth does not take place within the Valley, but elsewhere. The community suffers as income and employment decline, and a two-tiered economy provokes interclass stress (see SRI 1992: 59–66).

The creators of the scenarios pinpoint economic and educational infrastructure as key factors, while the vitality of the workforce defines the total success of the scenario. The prime mover, the perceived cause of the social change, is curiously one dimensional in these scenarios. The partnership clearly viewed the economics of high-technology industry as the chief architect of change, just as the medieval monastics once laid the locus of causation to God.

The original vision postulated by the community groups in 1992 has been periodically refreshed and re-created. In 1998 the report *Silicon Valley 2010* was published, the product of a major survey and focus group effort to collect and structure diverse opinions about the region's future. The executive summary is rich with cultural assumptions and values: "We will use our innovative, entrepreneurial spirit to create a strong foundation of regional stewardship, so future generations can enjoy Silicon Valley's broad prosperity, healthy and attractive environment, and inclusive communities" (Joint Venture: Silicon Valley Network 1998: 5). This version of the future sets four goals: stimulating an innovative economy; sustaining the local environment; including the broadest possible range of classes and cultures in education, transportation, and health; and creating a regional stewardship model that transcends traditional local political entities (ibid.: 7). Here again, continued growth in high-tech employment is a key assumption (ibid.: 18).

Beyond its central role in the economy, technology itself takes on a heroic role in the community. Techno-optimism—viewing the future through a rose-colored technological lens—can be clearly seen in two specific Joint Venture initiatives: Smart Valley and Smart Schools. One of the most striking examples of this perspective is expressed in the Smart Valley Initiative.

Smart Valley began during the economic downturn of 1992, implementing, in the words of Phillip, a Smart Valley board member, "a high-speed, fully capable, broad band infrastructure—so every home, every office will have access to high speed communications." Jerome, another member of the group, but an engineer, added that he believes that the "industry that was responsible for creating this technology" felt a concomitant responsibility for getting the local area to use it. Smart Valley had made marketing into a mission, using the language of a social movement. The ambitious vision postulated "the construction of a pervasive, high speed communications system and information services that will benefit all sectors of the community—education, health care, local government, business, and the home. The infrastructure we implement will help transform the way we work, live, and learn." With an annual budget of 1.3 million dollars, Smart Valley was the early Joint Venture's most heavily funded initiative, and the only one with a full-time president. Smart Valley also had a small paid staff, although most work was performed by volunteers and people on loan from their Silicon Valley employers.

The Smart Valley Initiative illustrates the union of techno-optimism and civic effort. Its goal was to create an "electronic community" by developing an infrastructure for information technologies and recruiting users for that technology. Its existence was intimately tied to its objectives. In 1997, as soon as it felt it had met those objectives, Smart Valley formally dissolved. The benefits of creating a networked infrastructure were assumed by Smart Valley's supporters to be widespread and significant. Smart Valley carried out its mission by supporting several initiatives of its own. These were selected on the basis of their consistency with the Smart Valley vision and their support by member organizations.

For example, the Smart Valley Telecommuting Project sought to enhance the capacity of companies to support their employees to work at least partially in their homes. Their rationale was simple:

With Silicon Valley businesses seeking innovative ways to maintain their competitive edge, recruit and retain key individuals and enhance the quality of life for all their employees, solutions such as telecommuting take on a much greater role than that of a "nice concept." The Smart Valley Telecommuting initiative is moving telecommuting from this "concept" to a recognized business strategy that provides benefits to Valley businesses, their employees, [and] to the region as a whole. (Smart Valley n.d.)

Another initiative, the Smart Valley Schools internet project, created a series of "NetDays" in which volunteer expertise was coupled with corporate donations to link K–12 schools to the internet. Apple, Cisco Systems, Sun Microsystems, and Intel were major sponsors (Tran 1996: 1B, 2B). In their scenarios of the future, our interviewees—whether they were educators or not—considered education to be in a pitiful state. Smart Schools advocates believed that networking of schools would "integrate technology as a tool to enhance the learning process and in the process teach students to live and work productively with technology."

The "efficient utilization of information technology" is a phrase that contains powerful assumptions about values. The power of technology, the superiority of efficiency, and the assumed dominance of a knowledge economy are imbedded in the designed solution. The vehicle for introducing that technology was an educational partnership with business. A survey conducted by the *Mercury News* indicated that 60 percent of Silicon Valley residents believe business should play more of a role in redesigning education. Businesses could provide equipment, guest speakers, mentors, and financial support. About half of the sample (41 percent of the nontechnical respondents) thought corporations in turn should expect a more technical curriculum, better exposure for their products, and input into curriculum or budget decisions (LaFleur 1998: 19A). Our interviewees—educators and engineers alike—echoed these sentiments in their scenarios of the future.

Industry academies were another attempt to redesign education. Business ventures were to be incorporated within schools to bring practical technical education to the students and aid in their school-to-work transition. South San Jose's Oak Grove High School program was an example of this blending of education and business in the "spirit of Silicon

Valley." The students worked with nonprofit organizations to diagnose Y2K problems and to develop e-commerce web sites for small businesses (Martinez 1999: 1B, 4B).

Three other initiatives were sponsored by Smart Valley. The Public Access Network provided free access to the internet and the worldwide web via terminals placed in public places. Bay Area Digital GeoResource developed a system for on-line visualization of geographic data about the San Francisco region. Bay Area Multimedia Technology Alliance worked to hasten the development of networked multimedia technologies and applications. All their efforts linked promoting technology with supporting civic endeavors.

Civic Engineering

Individuals use technological metaphors to talk about themselves— "upgrading" or "refreshing" their own skills, or "crashing" after a ninety-hour work week. The language of technology finds its way into civic discourse as well, drawing on obvious technological metaphors and tacit thought patterns. The organization Workforce Silicon Valley talked about needing to "reconnect the components of the educational system" (Sandoval 1996: 1C, 2C). Saxenian cites Silicon Valley as a "Protean place" able to "reinvent itself" (1994: 161). "Innovation" is no longer applied only to research and development but also to broader social processes (Delbecq 1994).

More subtly, Tech Museum and Smart Valley interviewees, whom we interviewed early in the decade, created scenarios of the future that revealed a particular way of thinking about technology and life. In most of the informants' scenarios, efficiency is the paramount value, and its consequences are seen to be enormous. Increased efficiency is celebrated in matters as diverse as buying groceries, getting to a destination, or cutting through government red tape. Phillip, a senior member of Smart Valley, explains:

The chief objective for Smart Valley is to make the electronic infrastructure for communications pervasive throughout Silicon Valley. This means that it's in all the schools, libraries, businesses, and governments. Because it will be within the infrastructure of society, it will enable society to operate more efficiently. Let's say that you have to renew your driver's license, and instead of having to go down to the DMV [Department of Motor Vehicles] you can send your information electronically to them. Think of it. All you have to do is electronically transmit a photo of yourself and then send them a digitized signature so they can verify that it's actually you. You'd never have to go to the DMV again.

The "platform" is another key metaphor for Silicon Valley informants. A platform is a technological foundation that supports many other devices or programs, just as a computer's operating system supports word processing, accounting, or browser programs. Most of these early informants described their homes as "platforms" for various media, and the number and integration of devices within the platform was clearly an important issue. Thus, educators see homes as platforms for student preparation. Homes are clearly described as refuges from a harsh and fast-paced business world. Work represents a threat to well-being because of the pace or amount of work expected, the ceaseless interruptions to daily routines, and the intense penetration of the corporation into individual lives. Home is, accordingly, a desirable place to be. On the other hand, isolated homes are also perceived as vaguely sinister. Individuals in them are, by definition, not in contact with others who need—or believe they have a right to—access. Individuals so isolated thus risk being bypassed by the flow of information and decisions that mark the successful information worker. And, perhaps even more threatening, such individuals might be unable to fulfill their social obligation "to be in touch," thus maintaining complex interdependencies. An isolated home is one that has failed to purchase the products and services that underlie participation in the electronic community. The implication for the home is clear. As platforms, homes must be networked to be effective. The new community is directly dependent on the metaphor of networking.

The 1990s also saw the birth of other institutions designed to help the community redefine itself, institutions that could transform the technological metaphor into a regional reality. One such is the Computer Museum History Center, still in the prototype and acquisitions phase. The other is the Tech Museum of Innovation. Initially called "the Garage," harkening back to the mythic beginnings of Hewlett-Packard and Apple's Jobs and Wozniak, the first Tech was a prototype museum, deliberately modeled on the entrepreneurial ideal. The Tech Museum of Innovation began with the idea that "Silicon Valley and the San Francisco Bay Area offer an opportunity to create an educational model linking industry, the schools and the community." The purpose is to "engage people of all ages and backgrounds in exploring and experiencing technologies affecting their lives, and to inspire the young to become innovators in the technologies of the future" (The Tech n.d.).

A core of advocates from industry and education developed a prototype "discovery learning" technology museum. They wanted to test the feasibility of an interactive venue that could inspire another generation of tinkerers. If it were successful in creating its educational product and attracting capital, it would expand. Tech staff developed exhibits that showcased the innovations of the Valley—new materials, semiconductors, robotics, and biomedical technologies. They developed outreach programs to regional schools and educators, varying from animation workshops for intermediate students to The Tech's Education Partners program for teachers.

Originally conceptualized as a place to which to send older elementary school–age children, it became a haven for families, a site for corporate celebrations, and a key destination for school field trips for both preschoolers and university students. Fathers were overheard explaining to their families that the gizmo for making chips is what they "work on." Children attempted to trick the "alphabot," a robot that stacked blocks programmed to spell out words entered into an attached computer, into writing something clever or profane. In addition to the physical structure, a virtual museum was built in cyberspace.

On Halloween day in 1998, the original prototype was replaced by a 112,000-square-foot center for science and technology education. That locally much-celebrated event inaugurated a "world-class" venue for the identification of the region with technology. At last, the region had a physical manifestation of its identity. Like Hollywood with its Universal Studios and Monterey its aquarium, Silicon Valley has the Tech Museum of Innovation. Much of what had made Silicon Valley renowned was corporate—inaccessible to the public and fairly dull to observe. The building of the Tech Museum changed that. It provided a place in which the handprints of Hewlett and Packard could be displayed along with state-of-the-art roller coaster simulations, IMAX screenings, and public internet stations. The museum provides a powerful technological metaphor for the region; it also provides a stage for the organizations that support and shape Silicon Valley. While by no means a blatant advertisement, corporate presence is manifest in the equipment, the volunteers, and the exhibit acknowledgments.

Charity is another vehicle for corporate presence. While much giving is not highly visible, expressed in small but persistent individual volunteer efforts, there is an organized philanthropic presence in the Valley. Compared with a national average of 69 percent, 83 percent of Silicon Valley households donate to charity; corporate philanthropy increased 49 percent between 1994 and 1997 (Joint Venture: Silicon Valley Network 1998: 2). Organizations such as the Volunteer Exchange match corporate talent with community need. Large corporations contain in-house units that manage philanthropy by bridging the human resources and public relations departments. These units scout out opportunities for philanthropy and match employee talent and corporate resources.

Between 1992 and 2000, more than a billion dollars was given to charitable foundations (ibid. 2001). Events such as the Silicon Valley Charity Ball are promoted and attended by high-profile public and corporate figures, and they raise millions of dollars for dozens of nonprofit community organizations: Adelante Mujer Hispaña, Boys and Girls Clubs of Santa Clara, Center for Living with Dying, Loaves and Fishes Family Kitchen, Los Amigos de la Biblioteca Latinoamericana, New Children's Shelter Fund, Saratoga Area Senior Coordinating Committee, Services for Brain Injury, the Tech Museum, and more. A foundation member commented: "We've asked companies to contribute $20,000 to our Million Maker category so we can guarantee raising at least $1 million next year, and the year after that. We're Silicon Valley. We should be able to do it" (Weimers 1998: 1B, 4B). Such charity functions form links, joining the powerful high-tech corporate world to the world of artists, educators, ethnic groups, and the disadvantaged classes.

What are the implications of reshaping civic life around the production of technology, and what metaphors accompany such production? In the last chapter, we saw the creation of a distinctive public space mediated by electronic devices. People work, interact, play, and create meaning, aided by various electronic media. This transforms their daily life in ways subtle and gross, redefining trust relationships and restructuring the definitions of work and home. The dominance of the high-tech political economy, and the absorption of technical metaphors into civic discourse, also redefine public culture.

The proximity of thousands of technological workplaces makes job mobility easier. People change jobs frequently and do so without the added complication of moving. Over time, the combination of organizational mobility with physical stability facilitates the creation of multiple linkages. Individuals link with other individuals, exchanging favors and strengthening the bonds between them. Networks link with other networks. Both individuals and networks interact with public and private organizations, blurring the boundaries between organizations and the public/private domains. These linkages are facilitated by technology, and take on metaphorical features of technology.

Silicon Valley culture is defined by being "innovative," "rational," and subject to conscious design. The culture views itself as doing significant social work, molding a future in which technology solves problems and facilitates communication. The importance that Silicon Valley attributes to itself becomes part of its "value-added" that can be used to attract the cognitive talent it needs to create its own future. The net result is that Silicon Valley itself becomes a commodity, allowing individuals, networks, and organizations to market themselves to the world.

However, as the people of Silicon Valley transform their civic culture into a commodity, they often forget that culture is also an agent in itself. The pervasive individualism of the region blinds people to the power of social institutions. Politics, education, and health care still matter, not just as "value-added" but as agents of change in themselves. Cultural identity is often viewed only as an accouterment or commodity. Hence, Hispanic workers are believed to help a company's products penetrate the Latin market. What is forgotten in this approach is that culture—as illustrated in differences in work values, family roles, even educational aspirations—is in itself a significant agent of change. Public life in Silicon Valley is partially designed "from the top down" but is also grown—an organic amalgam of ever-changing networks, companies, ethnic groups, migrants, and immigrants. How do the traditional markers of identity and identity politics—class, ethnicity, and national origin—play out in a civic culture dominated by high-tech economics and metaphors? To answer this question I must look at the second strand of Silicon Valley's double helix—the identity diversity and cultural complexity of the region. In the next three chapters I will explore the demographic consequences of living in a "silicon place." People, organizations, and cultural information flow through the region, picking up and depositing cultural bits. In this cultural flux, Silicon Valley people have developed a tool-kit of strategies for dealing with cultural difference. While some tactics are as old as classic ethnocentrism, others hint at an emerging ethos in which cultural homogeneity is not assumed, or even necessarily desired. The cultural dimensions of Silicon Valley are rarely directly linked with the technological hype that so often defines it, yet the two aspects, technology saturation and cultural complexity, are intertwined.

PART II

Trafficking in Complexity

CHAPTER 4

Input/Output

Emerging Global Culture

Relative Time

Lunch is over in Silicon Valley. Carl is back at work. His work is at the heart of the global interchanges that connect Silicon Valley with the rest of the world. He is the chief network architect for a company that builds networking equipment and provides services to a variety of clients. Their primary customers are American or European companies that work in India, where the networking infrastructure is problematical. Their investors are mostly American, but Indian money also supports the operation. To do his work Carl must understand the intricacies of the Indian telecommunications bureaucracy as well as the realities of routers. He has a meeting with an old friend who wants to know what he is doing these days. Carl's company is a start-up, so he is constantly absorbing information on potential customers and competitors to determine how his company should set prices and position products.

He conveys this to his inquiring friend, noting: "The vast majority of my job is data collection, information collection. I probably spend four hours a day on the internet, trolling. I spend a lot of time going through trade magazines, I go to trade shows, I interact with my peers from other companies, I farm for information from our overseas associates and whatnot. The primary job requirement is to be an information sponge." He makes a sucking sound, and trembling dramatically, pulls his hands toward his head. "It's kind of like the blue whale. You know, you go out there and screen through a billion gallons of sea water . . . to get a few brine shrimp."

After his friend leaves, Carl zaps an electronic message off to a contact in India. E-mail is the only way to go. He recently got a phone call from there in the middle of the night. Once in a while someone schedules a teleconference for midday India, forgetting to adjust for the time difference to California. He chuckles to himself as he remembers the incoherence of his response. He thinks that people are used to "absolute time . . . where the sun is [in the sky] for you." Carl is beginning to be comfortable with "relative time," time experienced both in the location of your body and the place you are contacting on the other side of the world. He understands it intellectually, but it is a different proposition to juggle the time zones in your head every day.

As Carl's message is being relayed to his contact, another message is being routed from South San Jose through Australia to Prem in an office in India. There, it is the middle of the night. Prem is an engineer, and he has an office space in Bangalore, India. His cubicle walls would look at home in any high-tech office in Silicon Valley, except for one fact: these walls were handmade by local craftspeople to resemble the mass-produced look of those in California. Prem's technology team in India is only part of a larger technology team, with five units distributed throughout the world. One unit is in the American Midwest; others lie in the heart of Silicon Valley. Together they are designing a component in a networking device that makes distributed teamwork like this possible. His team will make a small piece of technology that fits into a larger unit that will be delivered to the United States, and then find its way back to India when it is shipped to the Asian marketing division.

Walter, Prem's coworker in the United States, will call Prem from San Jose later that night. The phone call will clarify a design decision, sent via an e-mail message, that is waiting now in Prem's in-box. Walter and Prem can work together easily, since they can actually put faces to names. Prem has gone to the requisite meetings in San Jose, twice a year. Walter and Prem have had dinner together. They know each other's hobbies and family backgrounds. Their e-mail interchanges include not only technical attachments but also personal information, reinforcing an already established relationship of trust.

Waiting in Prem's voice mail is a message from his senior colleague, Mitesh, about a seminar being held that week. Prem has often had discussions with Mitesh about what they are doing for India while they work at their multinational high-tech company in Bangalore. Mitesh thinks that in his own way he is reshaping Bangalore. By holding seminars and encouraging open technical discussions, he is encouraging a "technology culture" like the one he has seen in Silicon Valley. Mitesh has been known to say, "I believe that innovation, which you need in the computer industry, can only come if there's a sort of atmosphere. For example, you have seen in Silicon Valley in the U.S., people are always talking about technologies and things like that."

Prem has relatives in Silicon Valley as well as colleagues. When he travels to San Jose on business, he also visits his sister in Fremont and makes new connections. Prem's brother-in-law is active in the Silicon Valley Indian Professional Association, and his network is most extensive. Prem harbors a quiet desire to work for an indigenous Indian company, not a huge multinational corporation. He would like to work on developing custom software. Maybe his brother-in-law's connections can help him to play the game.

There are many recruiters for the game. Jorge has just returned from the gym in his apartment complex. He sits in his living room in Santa Clara, the tools of his trade surrounding him. He has a laptop, an ISDN line, a phone, a cell phone, and the papers on his desk are arrayed before him. He is like a master puppeteer composing a play. His job is to find the talent to fuel global high-technology firms, and his specialty is Latin America. He must act as an interface between his "Germanic" Midwestern coworkers and the pool of technically literate, marketing-savvy Latin Americans.

As a head hunter, he must know how to match technical skills and cultural and linguistic competencies to available job niches. Jorge is uniquely suited to the task. Born in Argentina of European roots, he moved to California with his scientist parents. He draws on his own multicultural background to structure his arguments, judge when to bring up personal ties, and choose words and inflections in at least two languages. Conscious that the work day is now drawing to a close in Latin America, he picks up the phone. With each call, Jorge adjusts his cultural message, moving physically to a different part of the room, and mentally to a new cultural space. He checks his e-mail, each message color-coded by the part of the world from where it originates. Global economics is no longer the exclusive domain of the giant multinational corporation. Small startups must go global quickly to tap into the movement of capital, talent, equipment, information, and people that make up the high-tech world.

He chuckles to himself when he thinks of how he explains his work to his cousins in the Southern Hemisphere: "I am a pimp of sorts, and I buy and sell human beings. So a company wants a human being, they tell me what kind of human being they want. I go out and I search for this human being and I sell them back to the company and I make money that way." Some people in Silicon Valley import and export information. Others may import the architectural design of a piece of software and export the chunks of code that fill that design. Yet others may import money from Taiwan, turn it into product and export manufacturing work to be done in a foundry in Malaysia. Jorge's specialty is importing and networking the people and their everchanging bundles of skills.

People in Motion

Technology has a complex, intricate impact on cultural demography. Connected by technology, people move through the Valley virtually, exchanging information, products, and capital from around the planet. People visit Silicon Valley to learn and return, circulating technical expertise among the other "silicon places" from Boston to Taipei. People also immigrate to the region, adding their own attributes to the cultural mix. Technology provides the means—the logistical support—by which people connect to each other. It facilitates communication within the same building, or around the world. More subtly, technology reshapes social complexity by combining cultures into new mixes and providing alternative frameworks for reckoning identity. These twin processes of technological saturation and identity diversification merge in Silicon Valley.

People had moved through the region even before high-tech industry took root there. The Santa Clara Valley region was an agricultural magnet for immigrant workers long before its transformation into an icon of technology. The old and new communities coexist and intertwine. To make the cultural landscape more complex, it is not merely the immigration of new people but also the influx of capital, information, and interdependent products that globalizes the region. Flows, that is, "objects in motion," include not only people but also "ideas and ideologies," "goods, images and messages, technologies and techniques" (Appadurai 2000: 5). Consider the range of connections in the following illustrations:

—A product designer must understand the infrastructural constraints and working cultures of manufacturing facilities in Scotland, Ireland,

England, Singapore, Taiwan, and Japan, each of which produces components for his manufacturing process.

—A statistician manipulates marketing data from places as diverse as Southeast Asia and Russia, each site having its own ideas of what constitutes appropriate data, to create sales reports.

—A support person for Cisco must understand the subtle legal, cultural, and technical differences between Canada and the United States to provide customer service.

—An applications engineer works directly with customers and circuit designers in Silicon Valley, Germany, and Japan, through e-mail and face-to-face contact, to design and manufacture integrated circuits.

—Personal experiences with culture shock in South America have sensitized an instructional designer who now uses that knowledge to design culturally appropriate curricula and train trainers from Japan, Hong Kong, Australia, France, Germany, and Italy.

—A Nordic manager, with an Irish technical education and wife, uses his cross-cultural program translation experience to manage localization engineers who must adapt technology to culturally diverse users around the world.

Global sites are tied together in a complex web. People must work together. Goods and services must be delivered.

The complex interdependence is most pronounced in the interaction among the various "silicon places." A joke making the rounds is that Silicon Valley is the product of "ICs." The technically minded immediately know this refers to integrated circuits, but those in the know understand that it also means the "Indians and the Chinese." According to the 1990 census, 74 percent of the foreign-born technical workforce came from either an Indian or Chinese community, even if those communities were located in South Africa or Malaysia. Bangalore in India and the Taipei-Hsinchu corridor in Taiwan are products of interdependent development with compatriots in Silicon Valley (see Saxenian 1999). Ireland is a rising star in indigenous software development and is the site of transplants for such Silicon Valley companies as 3Com, Amdahl, Apple, Claris, Hewlett-Packard, Informix, Intel, Oracle, Quantum, Seagate, Sun Microsystems, Symantec, and Xilinx (Gillmor 1997: 1D).

Linkages can be complex. Large organizations can extend to many sites. Multinational giants IBM, Microsoft, Motorola, and Intel each have a presence in Israel, which contains yet another "silicon place."

"Two-legged companies" that have one foot in Silicon Valley and want another in a different "silicon place" use "matchmakers" such as the Bird Foundation or TechVentures to create links with countries like Israel, seeking to attract high-tech business. Global links can also be made when individual people go from one place to another. For example, personnel in Silicon Valley companies such as Adobe Systems or Applied Materials have left their jobs to employ their talents in Israeli companies (ibid. 1998: 21A). Silicon hopefuls such as Manila or Costa Rica court ethnic and business communities in the Silicon Valley to attract investment and talent (Jung 1996a: 1C; Dorgan 1997: 1D).

The job histories of our interviewees bear out the fluid links to other technologically identified communities. Austin, Boston, San Diego, Seattle, Edinburgh, and Taipei appear repeatedly in job histories and in personal future projections. Our interviewees know there are "peaks in other places as well. It's not just Silicon Valley. Silicon Valley is one of the highest peaks, but it's not the only peak." Once referred to as "brain drain," the flow of high-tech talent is now more accurately called "brain circulation" as people go back and forth (Saxenian 1999: 3).

The links connect more than individual people and organizations. Cultural brokers in "silicon places" link Hong Kong money with potential Silicon Valley investments. Irish intellectuals labor to adapt Silicon Valley products to a European market. Van, an internet executive I interviewed in Taipei, talked about how the products he developed were "based" in Silicon Valley to take advantage of American venture capital and legal infrastructure. Indeed, his company was planning to hire an established Silicon Valley chief executive officer to lend status and prestige to the organization, while the joy of creativity and research remained in Taiwan. Rising costs and restrictions on specialty immigration push the exportation of technical centers away from Silicon Valley, not only for manufacturing or localization of product but also also for research and development (Cha 1997: 16A).

Immigrant talent and joint ventures are not the only ways Silicon Valley culture is exported. A lively specialty tourist industry is being born. Foreign students, government officials, potential entrepreneurs, and journalists arrive in Silicon Valley daily to capture the "essence" of the Valley by visiting its high-profile companies and economic development officials (Quinn 1998). A Gray Line bus tour accommodates these visitors by taking them to Hewlett and Packard's garage and other sites of high-tech interest.

"Outposters" stay in the region in branch offices designed specifically to allow sojourners to be close to the heart of technological development. However, when questioned more closely, it is not the technology itself but the elusive culture of the Valley, with its fluid organizations and celebration of risk, that is being sought. Groups of managers and researchers from Ericsson or Daimler-Benz try to sift through Silicon Valley's alternative business plans and cultural artifacts, such as informal modes of address and casual clothing, to see if anything can be taken back to the mother country. Even when they are not seeking cultural transformation, these sojourners are exposed to it nonetheless. Some *chuzaiin*—Japanese workers on extended stay in Silicon Valley—although committed to their home companies began to consider the advantages that job mobility might offer for personal advancement once they were exposed to the Valley lifestyle (Teraguchi 1997).

The landscape of cultural pluralism reflects this high-tech international interdependence as well as the impact of previous generations of immigrants. The importation of people is a constant feature of regional culture. Kelly, a local, muses about the region's folk demography:

> In California a lot of Irish, being Irish, I'll say that first. Italians, some Germans, a little bit of Russian, English, Scots, and then Latino. Your farm workers, this was before Silicon Valley, this was when they called it Pleasant Valley, before it got nicknamed the [Prune] Pit and then it got renamed to the Silicon Valley. . . . At this point in my life I can look at somebody and say you're from here, immediately almost 99 percent of the time I'm right. And they go, "How do you know?" . . . When we were having these influxes of people coming, I got sensitive to each group of them. . . . A few people would come from New Jersey, and then they would go back and tell their friends, "Hey, this is great." Then we would get this "whop" of New Jersey people and then we got a "whop" of Chicago people, and they used to come in waves, from different sections.

The region's complexity, however, is even more dramatic than folk demography suggests. People are not simply divided among locals, sojourners, and immigrants. There is a complex history of movement in and out of Silicon Valley, making it difficult to track particular cultural groups. Unlike many regions, which can identify clear majority and minority populations, Silicon Valley contains a wide range of cultures.

In Flagstaff, Arizona, for example, the Navajo are the primary "oth-

er," and the distinction and coherence of the Native American community clearly identifies it. But a Navajo in Santa Clara County is merely
one "Indian" among people from many other tribes. Native American
students interviewed in San Jose view themselves as just one ethnicity
among many (Christie 1997), reducing the salience of their distinct tribal
identities. California has the highest Native American population in the
country, but within the context of cultural pluralism, the Native American population in Silicon Valley is a mere 0.7 percent (U.S. Census
2001). Any one group is but a fraction of a complex whole. This pattern
holds true for every culture in the Valley.

Figures from the California Department of Finance reflect the complexity of Santa Clara County's ethnic mix. While the 1970 figures
placed Americans of non-Hispanic European descent clearly in the majority, at 82 percent, by 1999 that group had become a minority of 49
percent. Hispanic-surnamed people constituted 24 percent, Asian and
Pacific Islanders 23 percent, and African-Americans 4 percent (Mc-
Laughlin and Cha 1999: 1A). These categories themselves are suspect,
however. In 1997 one of every seven babies born in Santa Clara County
had parents of different racial classifications, making the already problematic categories even less meaningful (Stocking 1999a: 1A).

The 1965 Hart-Cellar Act and, in turn, the Immigration and Nationality Act of 1990 dramatically changed the face of the region. The former
opened immigration from Asia as never before, and the latter tripled the
number of special occupation visas granted. A full 23 percent of Santa
Clara County is foreign-born, surpassing the proportion in San Francisco
(Saxenian 1999: 11). In 1990, an estimated 900,000 Asian/Pacific Islanders lived in the Bay Area. By 1999, the estimate had risen to 1.3 million (Akizuki 1999a: 1A, 10A). Forty-five thousand South Asians live in
Silicon Valley, and the number is growing (see Chen 1997a: 1A, 28A).
One-third of the sixty-three thousand self-identified Chinese in Santa
Clara County are foreign born: 36 percent from the People's Republic,
27 percent from Taiwan, 12 percent from Hong Kong, 14 percent from
Vietnam, and 11 percent from other ethnic Chinese communities around
the world (McLaughlin 1996: 22A).

These numbers, referring to gross categories of ethnicity or national
origin, vastly oversimplify the complexity of cultural pluralism. Ethnic
lumping masks disparities of culture, class, education, and attitudes
about work, gender, and family. A Vietnamese woman, probably one of

the 5,130 Asian women who compose a preferred hiring category for manufacturing, is worlds away from a Taiwanese man with a prestigious graduate degree from National Taiwan University and multiple venture connections across the Pacific (see Hossfeld 1988; Park 1996:163).

There are dramatic divisions among older and newer immigrant populations. When discussing ethnic communities, historians and social scientists separate longtime historic communities from contemporary emigrations. Consider the permutations of origins, affiliations, and statuses for one group—the ethnic Chinese. Chinese migrated to Singapore, Suriname, Indonesia, and San Francisco in the eighteenth and nineteenth centuries. In contrast, the Chinese in Vancouver, British Columbia, migrated in the late twentieth century, particularly from Hong Kong. These cohorts are fundamentally different. Silicon Valley includes both kinds of immigration.

Here there has been no one influx of immigrants, but many, combining the long-standing presence of a nineteenth-century Bay Area community with many waves of post-1965 immigration. This historic shift in immigration policy opened the United States to people from Asia, South America, and, to a lesser degree, Africa. For example, post-1965 Chinese immigrants are not a homogenous group, but come from China herself, Taiwan, Singapore, Hong Kong, and ethnic Chinese communities from Burma to the Philippines. These communities, once separated geographically, now interact through work, Chinese community events, and intermarriage.

Each wave of migrants from the home country brings a different version of the parent culture. A Cantonese-American from San Francisco whose ancestors came to California in 1880 has a memory of China characterized by Ghost months and Chinese New Year. His sense of Chineseness is different from that of the teenager from Hong Kong who views cell phones and Star television as an integral part of his sense of Chinese identity.

Immigrants bring with them memories that do not correspond to those of the nationals who stayed behind. The culture they came from has continued to change over the decades, or even centuries. In his book *Changing Identities*, Freeman discusses the sense of dislocation that Silicon Valley Vietnamese feel when they return to Vietnam (1996: 120–25). One informant noted, "I think Vietnam has changed from what I remember. Or maybe it is I who have changed" (ibid.: 124–25). Transnational sojourners, self-styled "astronauts" who fly home frequently, experience a very different slice of Ireland, China, or Vietnam than countrymen who stay in the United States and construct identities from collective memories.

The continuing population circulation makes creating simple categories of "Chinese" or "Irish" problematic. In the Chinese immigrant community, whether one is from Taiwan, Singapore, or the People's Republic shapes one's identity and social interaction accordingly. Others note whether someone is "FOB" ("fresh off the boat") or "ABC" ("American born Chinese"). As mentioned earlier, Silicon Valley's Chinese community embraces many different places, periods of migration, and sociopolitical orientations. Harry Wu, a leading dissident from the People's Republic who is famous for his adamant denunciation of Chinese labor camps and government corruption, resides in the area. But Silicon Valley also includes organizations of Chinese and non-Chinese alike who hope to build greater ties with China's new generation of technocrats, such as Jiang Zemin, president and general secretary of the Communist Party in the People's Republic (Morgan 1997: 6C).

There are many Chinese organizations within Silicon Valley. Since Chinese were once excluded from older "good old boy" networks, ethnic technical associations have proliferated, including the Chinese Institute of Engineers, the Asian American Manufacturers Association, the Chinese Software Professionals Association, the Chinese American Computer Corporation, the Monte Jade Science and Technology Association, the Silicon Valley Chinese Engineers Association, the Chinese American Semiconductor Professionals Association, the North America Taiwanese Engineers Association, the Chinese Information and Networking Association, the Chinese Internet Technology Association, and the North America Chinese Semiconductor Association (see Saxenian 1999: 29–30). Some organizations use Mandarin as the language of the group, excluding Cantonese speakers and linguistically challenged Chinese and non-Chinese professionals. Others use English and invite the widest possible circle into the potential network. The groups offer a range of services, varying from networking opportunities to English language workshops and workshops on how to write a business plan (ibid.: 30–36).

Often formed around a nucleus of former classmates and held in a venue that can provide the appropriate cuisine, the groups foster a link between ethnic and work networks, linking Silicon Valley to Asia. Sixty-nine Taiwanese companies make their home in Fremont alone (Akizuki

1999b: 1A, 22A). Similar groups have been organized for Indian, Iranian, Korean, Japanese, Israeli, French, and Singaporean technical immigrants (Saxenian 2000: 255). Sojourners from Northern Europe meet together as "Silicon Vikings."

The presence of many cultures and subcultures also generates new economic niches within the region. Specialized economic-ethnic niches abound. Chinese Saturday schools, cemeteries, corporate feng-shui consultants, and dozens of grocery stores cater to the Silicon Valley Chinese community. A local television station which serves the various ethnic Chinese boasts programs broadcast in Mandarin, Cantonese, and Vietnamese—as well as Indian, Japanese, Filipino, and Middle Eastern programming. An annual Chinese Festival at the San Jose Historical Museum draws tens of thousands of people, with mock archaeological digs and exhibits of ancient Chinese scientific achievements. Food is, of course, a central feature, as it is in all the many regional ethnic festivals.

The intricacy of the Latino community parallels that shown by the Chinese. "Latino" identity can refer to people from a wide range of national origins, from Argentineans who arrived last year to Californios with historic ties that extend back more than 150 years. It embraces Europeans from Portugal and Spain, and Native Americans who may be barely bilingual in a European language. Class and linguistic divisions are profound. Grandchildren of Mexican immigrants cannot really be lumped with people who came from Oaxaca last month. Fernando, an activist who works with a nonprofit organization, notes that while working in the Latino community, language is a key marker:

Chicanos are less concerned about Mexican culture or Spanish and are more interested in political involvement. Whereas people who call themselves Mexican or Mexican-American are more keeping their culture and heritage and language, and many are not concerned about being registered to vote or attending rallies but are strongly Mexican. Well, it depends on the individual, but Chicanos are definitely politically active and tend to be second- or third-generation Americans. I mean, go to any event which calls itself Chicano and you can bet it's in English. If you go to an event which calls itself Mexican or Mexican-American, you'll find a mix of Spanish and English or all Spanish.

Realistically, literary Spanish is not the linguistic marker in question. What is referred to here is a Californian creation—"Spanglish"—a local blend of Spanish and English. The local dialect is an amalgam that marks but one of the many Latino identities. Fernando points out that "parking," rather than the correct Spanish *estacionamiento*, is *el parkeo*. A typical Spanglish utterance might be: *Vamos a la playa man and we're gonna get all this food, pero también sería too cold so we're gonna need blankets pero mucha comida.*

Jorge, the high-tech recruiter introduced at the beginning of this chapter, exemplifies the class and cultural differences in the Latino community. The Spanish he speaks is Latin American, throaty and dialectically distinct from Fernando's Mexican Spanish. He is educated, a child of the professional elite, and few things irritate him more than being lumped into the same category as an "uneducated" migrant worker. His work, circulating talent, keeps him in constant communication with professionals in Latin America. They are the link to his identity, which he feels is more "Mediterranean," connected to the southern European immigrants to South America, than Latino.

Javier and his daughter Alicia illustrate yet another aspect of the life of Silicon Valley's Latino community. Javier was born in Mexico and spent his adolescence in rural Central California, classic Steinbeck country. He respects Chicano philosophy, but he is a "Mexican-American," valuing his "Latin roots" and "what this country has done for me." He is a successful manager in the world of high-tech human resources. His daughter is studying to become an engineer. Alicia had her first computer at seven and grew up attending conferences and professional network meetings. Her father proudly tells of the day he brought Alicia along when he went to speak to the Hispanic Business Association, and she stayed to network while he went on to another meeting. Javier notes, "She is going to be connecting with the Hispanic Engineering Association, and part of that is again *her* network, that *she* is developing." Fernando, Jorge, and Javier all illustrate some of the diversity and intricate interactions within the Latino community connected to high-technology economics.

Continuing transnational interaction among waves of immigrants makes it difficult to identify cultural affiliation. Gabriel is an American-born Filipino engineer who married a nurse from his ancestral homeland. Alena moved to Silicon Valley and went to work for Kaiser, a large HMO. Alena complains that the Americans she encounters don't speak proper English, and asks her husband to speak to her only in Tagalog. To Gabriel, colloquial American English is proper, and Tagalog is a foreign language. Gabriel thought he had married someone from his own culture, as his parents expected, but to his distress he has discovered that she does not really share his culture. But what is his culture?

When frequent communication with the "homeland" is added, the cultural landscape fractures even more. The relatives left behind are not so easily forgotten and molded into a distant memory when they are e-mailing the immigrant son and posting web pages. The establishment of transnational businesses—"two-legged" companies—links immigrant populations in Silicon Valley to those left behind. A dot com that links the transnational Indian music scene with internet savvy consumers via the technical expertise of Silicon Valley Indians is an example of this global cultural circulation.

The sheer number of possible daily interactions in Silicon Valley makes its cultural complexity qualitatively different from that of other places. In a place with three interacting cultures, the number of separate interactions is minimal. People from each category can interact within their own group, each culture can interact with one other, and all can interact. In total, seven interactions are possible. However, with fifty different cultures (using a conservative estimate), the number of potential interactions is well over 1.125 quadrillion. In addition, the gross categories of cultural identification used in the Valley are divisible into increasingly nuanced subcategories, further adding to the complexity of interactions. In the midst of such identity diversity, people cannot accurately predict the nature or outcome of their daily encounters.

Multiple Identities

Culture and identity are different concepts, but they overlap, making it difficult to carefully define the distinctions between them. Identity is one piece of culture. Identity refers less to what we do—our behavior—than to who we believe ourselves to be. Identity is reflexive, reflecting several dimensions; it is how we categorize ourselves to ourselves. Identity is theater; it is how we present ourselves to others. Finally, one aspect of identity is passive; it is how others classify us. Identification occurs at a number of levels simultaneously. I have one identity among my intimates and family, and another when I complete a census form. The tradition in American culture is to see identity through sociopolitical and psychological lenses. Americans have ethnic identities, or ones that stem from national origin, but Americans also view the world through "pop-psychologized" eyes—believing that each individual seeks out his or her own unique "self."

This customary way of viewing identity has two consequences. Tying identity to ethnicity or national origin overemphasizes that level of culture. We are led into the error of viewing culture as the same as "national character"—a deeply flawed and overgeneralized concept that leads to confusion when people do not act according to their assumed script. The overidentification of culture with nationality or ethnicity leaves out the other layers of culture—familial, associational, regional, transnational. On the other hand, when people think of identity as a purely individual choice they overlook the power of culture. Culture is a real factor in shaping how we think, act, behave, and create the objects around us, but contemporary American culture is not comfortable with that thought, preferring to view life as a series of individual choices free of the "shackles of tradition"—a phrase anthropologist Franz Boas used to describe the constraints of culture. Since culture does have agency—it makes a difference in how we think and act—the various cultures of Silicon Valley will influence work, family, and community life.

There is no simple configuration of ingredients that identifies us as a member of a particular culture; rather, culture is a vast network of objects, ideas, and relationships that can be viewed quite distinctly from other segments of the social structure. Identification, as a process, is a way in which people simplify the social structure to navigate its complexity. But the identity that emerges is only a feature of the culture; it is not the same as the culture itself. So rather than study one identity group in Silicon Valley, or "Silicon Valley Culture" as a particular identity group, we wanted to learn how people make sense of their own cultural identities. Do people identify with multiple categories? Are there patterns in the categories they use? How do those categories relate to each other? To understand the meaning given to different identities, it is necessary to ask, "What are the differences that make a difference?"

There are many alternatives for reckoning identity and culture. In our interviews we asked people, in several different ways, how they identified themselves. We observed the way material culture—assemblages of artifacts—was used to display identity. We watched as issues of "culture" were raised in discussions, public debate, and interpersonal interactions, which often masked struggles for identity. I based my interpretation of these observations on the idea that cultures—and by inference, ethnicities—are not biogenetic essences into which we are born, but complex social constructions that we learn and enact.

A classical anthropological example shows how many factors can shape identity. The older colonial political economy fostered a flow of people, creating cultural pluralism much as the new economy does today. The small country of Suriname lies on the northeast coast of South America. It was occupied by a variety of indigenous people, most notably the Arawak and Carib. The indigenous people were fishers and swidden horticulturalists, practicing "slash and burn" gardening of bitter manioc. In 1667, Suriname became a corporate domain that belonged to the City of Amsterdam and the Province of Zeeland. It was one of the first examples of corporate agriculture, and was economically viable only as long as an influx of African slaves could provide cheap labor. The Dutch had little cultural experience with slavery and were primarily motivated by an intense economic expediency. This resulted in an unusually oppressive system of slavery that led to marronage, the escape of slaves, in this case into the rain forest (see Van der Elst 1970), and by 1700 many Maroon communities composed of runaway African slaves had developed. The Maroons created a cultural mosaic that blended elements from diverse West African groups from Nigeria, Cameroon, Sierra Leone, Ghana, and Angola, incorporating indigenous American adaptations and aspects of plantation life (Hoogbergen 1990). Conflicts arose with the Dutch. Marauding bands of runaways so disrupted plantation life that the planters' society hired soldiers to eradicate them. These expeditions usually failed, strengthening the runaways and exacerbating the problems until the government found it expedient to grant them autonomy in 1760. Although the process was not free from conflict, eventually tribes of *vrije Boschneegers*—free Maroons, as differentiated from fugitives—emerged: the Djuka, Saramaka, and Matawai in the 1760s, the Paramaka, Kwinti, and Aluku-Boni (the latter largely in French Guiana) in the 1890s.

In the meantime, the plantations still required labor. East Indians, or "Hindustani," were actively recruited after the abolition of slavery in 1863. Indonesians—"Javanese"—were brought in as contract laborers in the 1880s. Other ethnic groups also came, including Europeans and North Americans, Portuguese Jews, Syrians, and Chinese from Hong Kong. Former African slaves became city Creoles. This colonially generated cultural pluralism is expressed today in distinct foods and languages, and is the basis of current political party organization (Dew 1990; Bureau of Inter-American Affairs 1998). Religion became a significant ethnic marker.

Ethnicity and race—that is, visible ethnicity—can channel identity.

One of the favorite games among the Maroon children was "the white man is going to catch you," a variant of tag and a threat parents use to make their children behave. But is race the only identity that matters? In spite of the obvious relevance of ethnicity in Surinamese daily life, a completely different structure of identification intersects and competes with it. When a new person arrives in a Kwinti village, people do not just assess his ethnicity but also note whether he was *baka busi* or *ana foto*—from the bush or the city. *That* is a distinction that matters. In part, it separates the Maroons and indigenous people from the urban "others," reinforcing ethnicity. But it also competes with ethnicity, forming a whole other axis of identification that cross-cuts racial or ethnic coherence.

Other identities also matter. Gender matters in this matrilineal society. Gardens belong to the women of a common descent group, and grant women a measure of economic and political power. Religious identity matters. Catholic Kwinti live in Bitagron, Protestants in Kaaimanston. Being a practitioner of *winti*, the syncretic African-Surinamese religion, is another key factor in defining identity (van der Elst and English-Lueck unpublished field notes 1977). *Winti* beliefs were a defining factor among the Maroon rebels of the 1986–92 civil insurgency (Thoden van Velzen 1990). Any attempt to understand identity politics in Suriname must be multidimensional, embracing not only the standard ethnic categories but also other social categories—spiritual, economic, and ecological.

Silicon Valley's complexity bears a resemblance to the colonially driven cultural pluralism of Suriname. In Silicon Valley culture, many elements are used to fashion identity. There are old and new ancestrally defined cultures, subdivided by language, gender, interaction with the "old country," and intermixture with other groups. Social and family roles form identity as well; singles, couples, and families with children form distinct groupings. Class also contributes to identity, though it may be hard to identify it as such, since class may cross-cut networks of friends and even families. Work is also a major factor in one's identity; workers identify with their work, their profession, and, sometimes, with the employer of the moment. Even which computer or internet service provider you use defines who you are, and more important, shapes who may be in your network. Each factor is not separate "culture," but each criterion provides a way for a person to conceptualize his or her identity.

The process of identification is based on a variety of criteria—ancestral culture, nationality, transnationality, ethnicity, gender, generation, class, occupational status, religion, and neighborhood. People talk about identity using different frameworks to construct their "selves." Many informants identify with psychoethical constructs, self-assessments of trustworthiness or work competence. Some identify with their professions: "I am my work." Others identify with work organizations, such as "Motorolans" or Apple employees who "bleed six colors," referring to the formerly multicolored corporate logo. Some identifications might be based on ethnicity, gender, sexual orientation, or family role (see Pickard 1997). Other sources of identity are less often articulated, but spring up as people create and resolve disputes across "cultural boundaries." These identities do not act as fixed categories but shift with different contexts.

This distinction was the basis of Fredrik Barth's ground-breaking thinking on the nature of ethnicity. He realized that people recast their identities depending on the audience, and he formulated a theory of ethnicity that does not emphasize unique fixed traits for each ethnic identity; instead, it draws our attention to the way people play with the boundaries of their identity when they are in a "dialogue with others" (Barth 1969; Cohen 1994: 120–22; Jones 1997: 72–85, quotation from Kuper 1999: 235). For example, Lowell, a process engineer, notes that he would identify himself with "Silicon Valley" to anyone outside of California. If he were talking to another Californian though, he would make it clear that he was from Santa Cruz—giving himself an altogether more "counter-cultural" identity, a distinction that would be lost on an outsider.

Silicon Valley contains many different categories of people, with finely nuanced shades of distinction. The distinctions that are meaningful to people within the group may be less obvious to those outside. Heterogeneous groups live side by side, unable to form sealed enclaves and forced to interact in schools, workplaces, and in civic struggles. The groups mingle in public places—while shopping, learning, or designing internet routers. Each group comes equipped with its own prejudices, assumptions, and cuisines. More profoundly, the groups are not stable. People intermarry and defy census categories by belonging to two or more groups. People adopt the cultural elements of other groups, and the very basis for identity shifts like the desert sands.

Viewing identity as a fluid process, rather than a stable symbolic category, changes the way it is studied. Instead of looking at an "essence," anthropologists look at how identity is expressed in interactions with others. Which cultural boundaries make a difference within categories? Do they reflect different categories, or different domains of life? Are these differences expressed verbally? Are we conscious of those differences? Do these differences alter our behavior as we act within and across boundaries? What is attributed to cultural boundaries, and what is not? This set of questions has led us to examine the articulation of identity, either by text—what people say—or by behavior—what people do—as individuals move among the different domains of life.

In a sort of folk anthropology, people typically construct their identity categories as lists of traits and collections of cultural ingredients. If you speak Spanish, have ancestors that come from Mexico, and your social life is centered around your family, then you *are* Mexican-American. But making identity a fixed noun—what one is—rather than emphasizing the process of identification—what one does and feels—ignores the many behavioral and symbolic markers that distinguish Mexican-American, Hispanic, Latino, and Chicano (see Legón 1995; Hurtado et al. 1993). Such labels conflate a fluctuating set of behaviors and attempt to freeze identity into an undifferentiated and static whole. This kind of labeling presents quandaries for people who do not conform to all the markers. Such a passive rather than constructive approach underlies the words of a woman who says she is not "a Chicana," since she rejects the trait "machismo." Labeling reflects the kind of fixed conception of identity that is demonstrated in a methodology that asks children to put their picture in a box with the "right" category to ascertain that child's identification (see Bernal et al. 1993). This fixed identity model reflects the way people *talk* about identities, but not necessarily the way they *act out* identities (see Cohen 1994: 120).

Global economics and population movements have made the region culturally plural. For many who work in high tech, the cultural variety is a heady experience. Working with people who do not share the same cultural assumptions about time, teamwork, boasting, or linguistic ability is a novel experience—and the differences are not merely ethnic. Tom, a manager, notes:

In this building we are truly a microcosm of what the next generation of American culture is going to look like. We have a whole bunch of languages spoken here. We have people here from a god awful number of countries and backgrounds because [we're] in a highly technical field. I mean we have Russian immigrants who are very good at mathematics and physics, as well as Chicanos who work at various levels throughout the organization, to people from just about every country in Asia, to white Anglo-Saxon Stanford graduates and University of Colorado graduates in a variety of different fields, to molecular biologists who come from the pharmaceutical belt on the East Coast who still are trying to figure out if California is for real. We have people like me who grew up in the South Bronx, one of the toughest neighborhoods in the United States, who also has a Ph.D. from Stanford.

In this utterance you can tease out the different factors that matter. While national origin and ethnicity are mentioned, so are class, regional identity, professional identity, technical prowess, and educational affiliation.

Identity markers not only go beyond ethnicity but also beyond these intraethnic distinctions. Many areas of daily life contribute to an individual's repertoire of identities, including gender identity, family structure, age cohort, network affiliation, regional and neighborhood identity, transnational identity, and class (see Mach 1993: 7). There are many foundations on which to build identification. Gender and class may cross-cut ancestral identities, giving a female Latino engineer common ground with a female Chinese architect. Former Apple programmers from Australia to Cupertino form an interacting identity-based group. Graduates of Landmark Education, an "individual and organizational effectiveness and communication" training, form enduring social ties around a common core of beliefs and activities. People from Palo Alto and the San Francisco Peninsula view their area as the center of their social universe and contrast themselves with people from San Jose.

People learn to emphasize one identity or another, depending on context. Fernando's story is revealing. Knowing that language is a marker of Latino identity, he considers switching: "I know this Puerto Rican guy and he doesn't feel comfortable with English and we got in the argument and I exploded in English and he said to me in Spanish, 'Why are you speaking in English?' and I said that I was sorry but it was a topic from work so I hear those words in English most often." He then reflects on the general rule, saying, "But it all depends, if you were to see me in a gay Latino night club then I use all Spanish. Because that's what's going on in there." Work, language, and sexual orientation interact with ethnicity to create situational identities.

This trilingual activist sometimes activates an ethnic identity, and at other times a gender identity. Fernando says, "For me it's situational. Like if I'm talking to my [Latino] brother I try to be pretty plain and masculine. Now when . . . everyone's hugging and kissing each other and teasing each other and being bitchy queens, I'm much more likely to play off that." Multiple identifications allow people to position themselves in context-dependent networks, navigating the social maze of the high-tech workplace.

Class is another factor that emerges in forming identity, one that is bound up with professional identity and the unique configuration of high-tech economies. Susan, a scientist and de facto engineer, adds, laughing: "I am probably a yuppie. I'd say you could define that as a kind of a culture." She pointed out that being a yuppie meant hiring help and focusing on work. People talk about the high salaries that high-tech employment can provide, which allow them to "outsource"—have others perform a variety of tasks, from housekeeping to arranging parties for visiting relatives. However, they also talk about the way high-tech work cycles leave them "time-poor" and make this same outsourcing a necessity.

Susan continued to reflect on her identity, elaborating on the class definition that supports her elite status:

I definitely identify with being British, and I definitely identify with being a scientist, and a physicist and you will notice that I keep calling myself a physicist even though my job title says engineer, because I've had people who are engineers tell me, "We can tell you're not an engineer by the way you think."

The class system in the Valley was repeatedly described not merely in terms of income but also as a two-tiered system of education, particularly by the people in the higher tier. Those with even a few years of higher education had one set of expectations and a particular sense of self, viewing those with less, or a different kind of education, as utterly different.

Intermingling distinct criteria for identity leads to wholly new forms—one engineer calls himself "the gay poster child of Hewlett-Packard." People also draw upon various interests and hobbies that are meaningful to them to shape their lives. Convinced that Nicole, a second-generation South American engineer, would expound on her national and ethnic heritage as defining her identity, I was surprised when she identified her culture as the one of "ultimate [Frisbee] players." I became sensitized to "dog people" and "cat people," or those who based their identity on bicycle commuting or motorcycle racing. A calendar depicting sailboats or Corvettes may contain a hint of an alternate self-conception that could be a source of consolation on work-weary days. Someone displaying such an image is not just an engineering drone but an adventurer who races cars or boats in his or her spare time.

In Silicon Valley it should come as no surprise that technology itself is a foundation for identity formation. Major distinctions emerge between users of different platforms—the distinct worlds of Apple, Windows, or Unix. Thus, even commodities play a role (see Cohen 1994: 177), as Mac and PC consumers find themselves on opposite sides of a great divide.

Linux, an alternative computer operating system, attracts true believers who express passion for their freeware and open source code. Fans expect that Linux will overthrow Bill Gates's Microsoft empire, democratize software development, and provide spiritual satisfaction to the user. Informants told us that a person might be identified as "a software guy" or a "hardware person," "much more than . . . because they are from different parts of the world."

The dominance of technology itself provides an alternative basis for identification in Silicon Valley. National origin and ethnic affiliation are only two of many foundations on which identity can be built. Work, the lodestone that has attracted so many Silicon Valley people, also provides an alternative basis for defining identity. In the local culture, work and worker identities dominate one's consciousness. For example, in discussing what to look for when hiring a person, a manager notes:

My perception of an Apple person is someone who is passionate, who believes in what they're gonna do, and has some focus around it. A non-Apple person would be someone who thinks that Apple would be a great name to put on their résumé and would be fun for a little while, but it's a stepping stone on the way to some other place. . . . So an Apple person would be someone who loves the technology, who is attracted to the platform, who is attracted to the people who use our equipment, because there is that whole creative thing that's what attracts a lot of Apple employees. We all feel that the platform has empowered us.

Gregory, a software engineer, views his role as a worker as the paramount feature in his identity, adding: "I don't think I put nearly as much time or energy into being gay or working with my gay identity as I do working for [the computer corporation]. . . . I choose to be at [the company]. I choose to be a part of this tribe." Given that people spend so much of their waking time at work—sixty to one hundred hours a week—it is not surprising that Erin adds, "I think I belong to different cultures. I think there is a work culture that I belong to, that is very get-things-done. . . . There are a lot of very intelligent people that work in this building, that have the drive and the motivation to do a lot and get a lot of things done, and push. I think that's a culture that I . . . get caught up in."

Identifying Cultural Agents

Culture organizes ideas, actions, and artifacts that we associate with that most domesticated of species—human beings. Culture transforms our environments into artificial spaces. Culture provides the many and often contradictory rules that shape our relationships with other human beings—how we act, what we tend to think is important, and whether or not it is acceptable to eat insects or design microprocessors. It shapes emotions and actions.

When people use the word "culture" in everyday language, it has come to mean several things. If you hear the phrase "It is just cultural" in a business meeting, it may mean that the behavior in question is a custom, part of a traditional way of doing things that surely cannot survive in the face of the better way—"my way"—of doing things. It is an ethnocentric dismissal of the exotic (see Kuper 1999: 10). It has also become a euphemized code word for race, conflating "cultural" identity with "racial heritage" in sanitized terms that seem "less racist" (ibid.: 240–41). "Culture" is also a term that is used to explain the unexplainable. People use the word to explain a behavior that mysteriously varies from a person's expectations. If the action was perplexing, "culture" must be the culprit. This is analogous to the old saw in archaeology: "If we can't figure out its use, it must be religious." An archaeologist finds material objects; religious beliefs cannot be "excavated" but must be inferred. Similarly, since culture is not usually articulated consciously, people must guess about the "cultural origins" of a particular behavior.

Culture is a social agent—a factor that makes a difference. A passion for Peets or Starbucks coffee over Folgers seems like a random individual choice until the collective behavior of markets and consumers is examined on a global scale. Methods of coffee production and distribution underlie changes in the market. Until valve packing ensured a fresher product, and new transportation technologies allowed whole beans to be shipped to distributors and consumers, custom coffees were not widely available. The association of premium coffee with class and generational status changed the profile of the whole industry (Roseberry 1996). Drinking "yuppie" coffees has become an identity marker, but it is also embedded in broader forces. We must look for culture at the intersection of complex forces—technological, economic, organizational, and ideological—not just in the acts of individuals or identity-based groups.

One way to identify the role of culture is to ask, when does culture offer a genuinely different set of alternatives? When does culture make a difference in what is happening in the community? In the following examples, culture is an active agent in influencing behavior. It shapes attitudes about diversity itself. It influences what we think of as private—not to be disclosed to others—and what we think is public. It shapes the way we think of the elderly, and what constitutes "respect" for the elderly. Family structure and family obligations are similarly culturally constructed. We see the agency of culture in the following examples.

One aspect of Bay Area culture is its familiarity and level of comfort with the landscape of diversity. People talk about feeling uncomfortable when they encounter a culturally homogenous social scene elsewhere. Carol, a process engineer at a major company, comments that when she is traveling, "I'm so aware of when there is not diversity, it just feels uncomfortable and odd to me . . . this is really odd, all this sameness." Kristin, a K–12 teacher, having traveled to other, less diverse places in California, suggests:

I remember coming back here and noticing that there were different cultures and different languages. That even in the grocery stores, in the food items that you want there is more diversity. I like it. . . . I've lived in the Bay Area . . . my whole life. A Bay Area, Northern Californian culture . . . things are fast paced . . . high-tech, busy, and competitive, and that's the culture that I identify with. And they are diverse, and that's one of the reasons I like this area and culture, if you want to call it that, because it is so diverse.

One of my anthropology students, trying to puzzle through the nature of Bay Area culture, seized upon the elegant, and obvious, notion that the presence of many cultures *is* the local culture. Rachel, a Berkeley native, elaborates on the impact of living in a complex cultural landscape:

It's probably so much a part of my culture that I don't even think about it. . . . Berkeley is a part of my culture and growing up in the Bay Area, what does that mean? It's liberal, it's tolerant, it's diverse, actually, truly in ways in which, you know, you go to school and live in neighborhoods with people who are not like you. They are not white-American-Jewish. I mean you may meet some white-American-Jewish people, but you're also gonna meet some black-American-

Muslim people or some Japanese-American-Buddhist people. So it's . . . exciting. I think the Bay Area's a really exciting place.

This adds to the picture of diversity Kristin paints: "Well, in the classroom I've got . . . kids from India, China, Korea, Mexico. I've got kids from all over." While the implications of living in this diversity will be discussed more thoroughly in Chapter 5, the very experience of living in this milieu can become the basis of identity. Northern Californian culture can become an identity in itself.

The Bay Area "experience-the-world" society opens up a range of cultural artifacts from which people can pick and choose. Kyle, a painting contractor, notes that cultural diversity "affects me in the sense that I am able to experience different cultures by their languages, their customs, . . . how they celebrate certain holidays . . . and that affects me in a good way." Just as food differentiates the regional identities of China—no one could conflate the fiery flavors of Sichuan with the subtle sauces of Shanghai—Bay Area identity is marked by the range of cuisines and the importance of good food. The painter adds:

I get to experience different foods. In other areas, like if you went to Japan and had a burrito it would be made by a Japanese, but here I could have a burrito made authentically by a Mexican person here—maybe with a recipe that was passed down through his family over the years. I'm not talking Taco Bell. I could eat Russian food here made by Russian people and so on and so forth. Indian food made by people from India that is authentic in my mind, and that is a neat thing to experience.

Locally, ethnic cuisines are not necessarily "authentic," or as they were in the "old country," but have been adjusted to the Californian palate. Yet Kyle views the many forms of food as markers of the deep diversity of the region. He adds some provisos, as if realizing that the benefit of global cuisine carries a cost: "Sometimes it is difficult to understand their customs, as a businessman, and bidding work, their procedures of taking shoes off before you come in or sometimes if you don't know certain things as far as negotiations, such as Chinese negotiate differently and it can be a challenge." The food *and the confusion* are both part of the culture that the locals identify as their own and that the newcomers view as part of the Silicon Valley experience that is shared with the wider Bay Area.

In Silicon Valley, "diversity," like "community" and "family," is a powerful symbol whose meaning varies with each speaker, although Silicon Valley denizens act as if they all mean the same thing by the word. Joint Venture: Silicon Valley Network, a consortium of business and community elite, often uses the word "diversity" to express a range of economic niches and professional specializations. Yet when these words were uttered in a public showcase of Joint Venture proposals for community changes, they were voiced by a panel of selected "people of color," suggesting a very different meaning.

Tom, a senior manager in a large research and development division, elaborates on how this diversity is seen in the workplace, noting:

We have an incredibly varied culture. I have a suspicion that this building may be the most culturally and intellectually diverse in the corporation. The reason I say that is that we have the classic electrical and mechanical engineers, classic marketing type people, and we also have a huge number of chemists, molecular biologists, and biochemists. And a very large, diverse software staff.

Distinctions among these professional affiliations are meaningful. In the same company, a marketing specialist adds with a degree of wry humor:

I think . . . one of the big contrasts [we have is] between the people who have worked in the biological sciences where nothing is certain, and the people who have worked in the engineering sciences where if it's not certain, then it's because you're an idiot and you don't know.

People struggle with the consequences of job mobility where they not only need to be able to work within multiple machine environments but also to know the difference between Adobe and Cadence corporate cultures—how decisions are made and resources acquired, and whether or not the coffee is free. Justin, a worker from a networking design company, tells us:

When we came [here] it was very apparent, and this was made very clear to us, that [this company] has no . . . technical religion. Their only religion is the customer. So if the customer is gonna buy ATM . . . we are gonna build ATM. We're also gonna build for Ethernet and any other darned thing that the customer wants to buy because we, as a company, are interested in customer success and in making money.

He laughs, adding: "And you know, whether that offends any of you pure and clean engineers that don't want to work on this dirty technology, well tough luck because [this company] has no technical religion." He is talking about a kind of diversity, centered around technological preferences, that is as much a part of the cultural landscape as multiculturalism.

Members of different cultures have different expectations about disclosing mistakes—a vital part of design and testing in high-tech work. Ari is a young Israeli manager of an engineering group in one of Silicon Valley's high-tech companies. One of his objectives is to establish a testing group that tests the product and reports back to the development group so that they can correct any flaws. Chen, a Taiwanese engineer, had just finished the development of their first product. When the test group's report is returned, Hsieh, a senior engineer in Chen's group, reacts furiously, screaming, "How dare you say all these things about my product!" Ari tries to soothe Hsieh, making it clear that blame is not the issue; they are only discovering problems in order to improve the product. It takes months for the feelings to settle down.

Ari's team tries to install a database system that will report problems during the development process. This effort is resisted by Chen's team. Chen feels that such a system would allow everybody to see his group's mistakes, and that puts undue pressure on them. Ari again explains that problems are expected to appear. Over the next few months Chen's group does not use the system, but works directly with the developers instead to fix problems without reporting them. From Chen's point of view, the expectation is to keep one's "error profile" low and informal. This contrasts with Ari's high-tech corporate managerial style—fast-paced, high profile, and very public. The different operating styles heighten corporate tensions.

Cultural differences are felt within the private world of the family, as well as the public world of the workplace. For example, consider the range of attitudes toward the aged in various cultures. This cultural factor shows up when immigrant populations, from cultures that emphasize gerontocracy, where power is in the hands of the elders, encounter cultures that do not venerate older people. Problems occur when the oldest generation of an immigrant population must look to the younger generation to learn about the new culture, overturning the traditional rationalization for giving power to the elderly—because they know more. In effect, each generation is partaking of a different set of cultural expectations.

Louis, a Taiwanese man who has lived in the United States since he was a teenager, works in his family print shop. At work Louis and his mother verbally fight, loudly and angrily. Louis wants to upgrade some equipment; she insists it is too costly. She throws things at her son, while he screams at her. Louis speaks in English, his mother in Chinese. She then demands an apology. Once, an earthquake shook items from the shelves and the lights overhead swayed, but the old woman would not move until her son apologized. In another incident, Louis anglicized the pronunciation of his family name to relate better to English-speaking customers, and his parents became furious. While identity is an issue—whether Louis is Chinese or Chinese-American—the central conflict is over cultural differences about age and authority.

Culture can be an issue in families as well. Mary is a twenty-six-year-old Chinese-American woman who immigrated to the United States from the People's Republic when she was eight years old. Her family is quite Americanized. When she was twenty-one she married her husband, the youngest sibling in his family and a recent immigrant from Taiwan. Gene's parents insisted that she live with them after the wedding, but she wanted her own home, much to their anger and distress. Finally she gave in and moved in with them. Her in-laws plan her weekends for her and do not allow her to express her opinions. She is expected to serve them at dinner and do the dishes, and she does so, although unwillingly. Both parties are insulted by the perceived lack of respect they receive. It is not just identity that is at issue but also the culturally constructed ideas about what constitutes proper role behavior.

While these examples are drawn from immigrants—each generation illustrating different cultural expectations—I want to draw your attention to the power of culture in fashioning those expectations. If culture can be a powerful agent across such nuanced differences, what can it do when greater cultural diversity must be negotiated?

Troubleshooting Culture

In Silicon Valley, culture is confusing. Sometimes culture sets up a smokescreen that makes it hard to understand other potential causes for behavior. If a software program works on one platform—Windows—but does not work in a Unix environment, a software engineer must "troubleshoot" the cause of the dysfunction. That is a particularly difficult task, involving a huge number of variables. Similarly, it is not always easy to tease out the "cultural" dimensions of an encounter from the maze of conflicting corporate policies, personal styles, and political positions.

Consider the following story: Hua is a middle-aged ethnic Chinese immigrant from Burma. A decade ago, she moved with her family to the Bay Area. After she spent a grueling stint in the Cantonese garment factories of San Francisco, her daughter found her a job in manufacturing at a high-tech corporation in Silicon Valley. There Hua has steadily cultivated her skills and now supervises a small group of workers, largely female. In the research and development facility where she works, production line employees take a daily break in an informal mini-cafeteria. There she may chat with other workers. However, when she uses Mandarin to speak with some coworkers, others chide her, citing the company's "English-only" policy.

Hua smiles and notes that no such policy exists, particularly for break times, but she understands that her coworkers are nervous about not knowing what she is saying, and afraid that she might be criticizing them. In an organizational milieu rendered somewhat paranoid by reorganizations and downsizings, people are nervous that someone in a position of even modest authority might be critical. This anxiety is compounded when potentially relevant information is spoken in a different language. In this case Hua's Burmese-Chinese culture does not matter except as a symbol of informational inaccessibility and loss of control—she could be a Navajo speaker and it would not make a difference.

It is easy to attribute the cause of a problem to culture. People make snap judgments about other people's motivations—often framing their understanding of the other person's behavior in national or ethnic culture. This inaccurate appraisal might be the result of other causes entirely, such as class conflict or personality clash. When that happens at the community level, difficult to resolve social conflicts arise. In the city of Cupertino, tensions between non-Asians and the Chinese community surfaced in public debate about a series of issues. Cupertino is a city of fifty-one thousand people in the heart of Apple computer country that has been transformed by the presence of Chinese immigrants. Drawn by jobs and the stellar reputation of the school district, the Asian population swelled from 7 percent in 1980 to 28 percent by 1998. Chinese language signs and books in Mandarin have penetrated the formerly largely homogenous Euro-American community. A proposal for a two-way Mandarin language immersion program in the public schools generated passion on both sides. Objecting that Mandarin instruction should be done at home or in special schools, some non-Chinese viewed the idea of fostering such learning in public schools as an "imposition." Rising housing costs and the increasing number of "pink palaces"—huge residential architectural monstrosities—are attributed to the Chinese, and their "exotic" extended family patterns. The obvious education and wealth of the new immigrants is offensive to the older, working-class residents (Stocking 1999b: 1A, 14A).

Although genuine cultural differences do exist—such as differing attitudes about education—"culture" in this case is largely a smokescreen, obscuring the more difficult class differences and tensions produced by economic change. When culture provides an easy explanation for mysterious differences, people fall prey to the trap of attribution, and the stage is set for stereotyping and intercultural hostility.

In a complex community, where identities intertwine and people of different cultures interact, it is impossible to assume that another person shares your approach to life. Cultural expectations are assaulted every day. Yet the work must go on. What strategies have Silicon Valley people developed to manage this complex identity diversity? How does cultural complexity—and the tools used to manage it—interact with technological saturation? Irene, a diversity consultant at a high-tech company, coins a metaphor—typically scientific—to describe the people around her. To justify to her company the utility of acknowledging cultural and social differences, she invokes the idea of absolute value—the distance of an integer from zero on a number line. One can have a negative number such as -3, to the left of zero, or a positive number of the same value, 3, to the right of zero. The absolute value of negative three, $|3|$, is always just three, since the absolute value measures distance from zero, and distance is never negative. As she notes, "Those absolute value bars that you put around negative 3, the value is still three." Then, using this analogy to describe how people must treat each other to work together, she adds that people must "put the absolute value on what's there," rather than emphasizing the details of the difference, the distance from zero. She adds, "It is a really great concept!" Rather than highlighting the "direction" of ascribed gender or cultural status, she advocates focusing on the "absolute value" of their achieved worth to the organization and the community. This is but one of the many strategies used to manage difference—some old, others more recent inventions. In the next chapter, I will examine these strategies, and the underlying principle that drives their use—cultural instrumentality. Thus, Silicon Valley's people transform their cultural knowledge into a powerful tool.

CHAPTER 5

Executing

Culture at Work and Home

Cultures of Care

It is late afternoon in Silicon Valley. The traffic is beginning to thicken dramatically at the intersection of highways 280 and 87 near the downtown headquarters of Adobe software. The Cupertino stretches of highway 85 slow to a fast crawl as the first wave of the flex crowd from Apple and De Anza College hit the highway. In Miner Park, nestled in South San Jose, a group of nannies are watching their children after their afternoon naps. The group is as multicultural as any team of engineers drawn from the global pool of talent. Angie is a young African-American woman. Bhavandeep is an ethnic Indian from Fiji. Yolanda and Ruby are from Mexico. Liddie, a local, is a nanny-for-the-moment, while she does course work toward a teaching credential. Yolanda and Ruby often meet at the park to watch their charges and discuss Spanish-language soap operas in Spanish. Today the group is larger and more diverse. They are talking, in English, about children's television, the weather, and their favorite topic—the families that employ them.

The women find themselves in a strange position. They take care of the children, spend time with them, even love them, but they know that they are, in the final count, employees. Alisdair, a high-tech quality assurance manager, often calls Bhavandeep "family." Yet when a guest comes to the house, he doesn't bother to introduce her. She has been with her "family" for three and a half years, living in their house Monday through Friday. When Alisdair and his wife go out of town, Bhavandeep will stay over the weekend as well. She does more than watch the children. She makes the beds and does the washing and ironing. On the weekends, she goes to her daughter's house. Although she doesn't exactly put it in those words, she understands that her employers are from the British Isles, and cultural patterns of servitude are different than they are for her friends' employers. Are Americans different in this regard?

Ruby laughs. While her work is identical to Bhavandeep's, her life is not as separate from the lives of Michelle and Roy, her employers. She is thirty-five and has two sons of her own, now in high school. She tried being a temp office worker when she was younger, but she didn't like it. She missed her children and felt like people at work didn't treat her with the respect she deserved. Her new employers don't mind if she brings their girls to her house, where she can be with her own kids. Her sons have grown to think of the girls she watches almost as their own sisters. They are teaching them Spanish. Ruby's husband, Jesus, also loves the girls she looks after, but he worries that Ruby has invested too much love in them. He understands that the girls are not his, and that eventually they will grow up and become aware of the great gulf that divides them. This realization makes his heart heavy, which in turn disturbs Ruby.

Even so, they are two families locked together in a tight orbit. Ruby's youngest son has fun trying to teach the girls to dance. The girls' parents even help Ruby's oldest son out sometimes with his computer class. Roy, the girls' father, and Ruby's son chat about web sites and such. Ruby really doesn't pay much attention to all that. Sometimes she will stay later at the girls' house and cook with Michelle. They exchange their homemade jams and chat. At times, Ruby and Jesus wonder how Michelle and Roy can stand being separated so much from their little ones, but they respect the hard work they do.

Yolanda is frankly perplexed by her employer's behavior. Melina telecommutes, and when she is at home, she is at work. Sometimes when Yolanda returns from the park, she will take the little one to her mother to say "hi." But Melina discourages too much interaction, telling Yolanda that it is her job to make sure that Melina can do her job. Even if they are both at home, Yolanda cannot ask Melina to watch the child for a minute without a sense that she is stepping over a line.

Liddie has a clearer idea of her employer's expectations, and tries to explain to Yolanda why Melina puts work first. As a future school teacher, she is more educated than the other women in the park, although Angie went to college for a few years. Liddie's employer has told her she wants her to instill the value of education and hard work into her charges. In her student teaching experiences, Liddie has seen that putting learning first allows the many diverse students to work together despite their quite different backgrounds.

Angie knows she is needed by her employer, Sandra. Like Bhavandeep's Alisdair, Sandra calls Angie part of her extended family. Without Angie, Sandra certainly couldn't concentrate on her work as a high-tech marketing manager, nor could her husband, Geoffrey, focus on his work as an engineer. When Sandra and Geoffrey decided to have a third child, they knew they had to make a choice. Sandra could stay at home, and be bored out of her mind—and they would be much, much poorer. Or she could get a nanny and accept the promotion dangling in front of her. Angie has worked with young children "forever," and Sandra feels fortunate to have her as her nanny. Angie enjoys playing with the children, inventing games and taking them on endless "field trips." Angie's mother is an educator, with a doctorate in education, so she spent her childhood with mom and her siblings playing "educational games." She grew up with a strong sense of her African-American heritage, expressed in church, in family, the food she cooks, and the music she likes.

Melina and Michelle, Yolanda's and Ruby's employers, have expressed delight that their children are exposed to a different language and culture. But what does this mean to the children? Crystal, Michelle's older girl, chats with Ruby. She knows that she and Ruby live in different worlds. There is the world of Mom and Dad—the trips to the Sierra, weekend visits to the Monterey Aquarium, the games in the backyard, the nighttime stories. They use quiet voices inside. It's different when she is with Ruby and Jesus. They play loud and boisterous games with the boys. Ruby teaches them Christian hymns. They laugh a lot at Ruby's house. Jesus hoots at the television during football games.

At Ruby's house, Crystal watches a few minutes of Nick Jr., cable television programming for young children, and then Ruby's son puts in a Disney video. Ruby begins to cook supper. Crystal loves television. There is no television at her preschool, where the emphasis is on education. Crystal really loves her preschool teachers. The day before, Lucy, her Latina teacher, had showed Crystal how to play "dress-up" and "princess." Teacher Lucy asked, if Crystal was the princess could she, Lucy, be the queen? Crystal told her she could be the "housekeeper."

Switching Windows

In the last chapter, we saw that Silicon Valley contains a vast number of cultural interactions—at school, at work, and in the hearts of courting couples. Sometimes interactions involve obvious cultural differences taking place across a barrier of language or class. At other times, people experience differences more subtly and are confused by a sense that they "should" belong to the same group, and have a common set of assumptions. In cross-cultural interactions, there is more than one set of rules for how people should behave. Which should be in the foreground—differences or similarities? If they emphasize their differences, will there be a danger that one identity will dominate and overwhelm the other? If the people in the interaction minimize their differences, placing their separate cultural identities in the background, what feature will define their interaction? The choice of which persona to put in the foreground again depends upon the context—the people, their cultures, and their tasks. In Silicon Valley, many interactional permutations are possible, and often several interactional styles are taking place simultaneously. There is no one interactional style. Yet we can map the permutations and note their consequences for individuals and for the community.

Silicon Valley's identity diversity embodies the dilemma of the "new ethnocentrism." As you recall from the first chapter, such cosmopolitan ethnocentrism occurs where many cultures coexist in a social setting, and it is difficult for people to assume commonality or even accurately assess one's own or another's identity. Clifford Geertz, in his reflections on the future of anthropology and humanity, marks this as one of the conceptual challenges of the twenty-first century (2000: 86). He comments that what is needed are "ways of thinking that are responsive to . . . 'deep diversity'" (ibid.: 224). If people experience social interactions in which there are many ways to "belong" or "be," they are living in a state of "deep diversity." The "new ethnocentrism" is the other side of deep diversity.

Classic ethnocentrism is built into being human. "In the process of learning culture, people come to regard the particular version of it that they learn—their own culture—as part of the natural world" (Bohannan 1995: 21). This sense of cultural entitlement, that our own practices are sensible and proper, forms the basis for a visceral rejection of other cultural possibilities. Ethnocentrism, in its classical form, produces a strong sense of cultural "self" and "other." People can confidently identify "us" and "them." However, in a complex society in which people are exposed daily to multiple versions of cultural values and assumptions, ethnocentrism is also more complex. "Us" and "them" grow harder to define. To a male Midwestern engineer, is an engineer from Bangalore the "other?" Or is that biologist a more obvious "other?" What about a female British engineer who was trained in physics? If the Indian engineer went to MIT, the Wisconsin engineer's alma mater, is he less or more "other"? Is the American programmer, twenty years his junior and covered with body piercings, more or less alien than the Indian?

The new ethnocentrism reflects ambiguity, not just rejection. Rather than simply rebuffing "the other," people are uncertain just what they should do. They do not necessarily know how their own cultural identities will interact with others, since all the identities are in a state of flux. In a society with more than a quadrillion potential intercultural interactions, how do people know how to behave with each other? Ambiguity and uncertainty are the hallmarks of postmodern ethnocentrism. Silicon Valley, with its flow of global cultures and alternative forms of identity, is a natural laboratory for the study of deep diversity and the new ethnocentrism.

In Silicon Valley, technological devices of every description shape distinctive cultural processes. There is a fine irony in the way that cultural identity itself has become a device—a tool for connecting and disconnecting people in a network of interactions. As I discussed in Chapter 4, Silicon Valley denizens draw on many identities and cultural artifacts. Identities that are apparently simple break into many fragments.

Mari embodies the intricacy of Silicon Valley's cultural landscape. Mari is Japanese-American. Her parents recently converted from Christianity to Buddhism, joining an Asian-American religious community. They object to Mari's intimate—but unmarried—relationship with Josh, with whom she lives. Mari met and became engaged to Josh, a native of Santa Cruz, a community noted for its "hippie" counterculture. Both Mari and Josh are vegans—they eat no animal products, not even honey. Their dietary preference is an important cultural marker for them—part of their identity—and their shared lifestyle draws them together. Yet her parents object to him. His ethnicity is not an issue—six of her relatives are in "mixed marriages." Mari's parents feel that the young people are violating the norms of public and proper display of intimacy by living together while unmarried. Following the dictates of her own conscience, Mari continues to live with Josh. Her parents are dismayed but resigned. Mari has no simple guidepost for propriety, but must draw on a wealth of frameworks—ethnic, filial, religious, philosophical, and romantic—to negotiate her way through the complexity of her relationships. She is living in a state of "deep diversity."

All identities are not equal, however. Racial identity, linguistic competency, political power, and class distinctions are still very real markers of status inequality. Despite a rhetoric that claims that in a techno-democracy, a Latino shipping clerk can e-mail a criticism to a WASP CEO, most workers recognize that to do so would be to commit employment suicide. Power is embedded in the infrastructure, and mechanisms for social control are still in place. Intercultural negotiation is weighted with realpolitik, but made more ambiguous by the abundance of cultural choices. People use different strategies with those they can determine are "weaker." The difficulty remains, however, that statuses can change radically in a technical meritocracy, and unlikely as it may be, today's flunky might be tomorrow's CEO.

In this socially uncertain environment, how do people assess and project the elements of their identity? Differences in ethnicity or gender might be the primary focus in a particular encounter, or they can be pushed to the background. Being an electrical engineer, or a circuit board designer, or a Christian might prove a more salient identity to embody and project. Like a computer running multiple programs in different "windows," one screen is in the foreground, while the others are running in the background. The choice of which screen is dominant depends on which task is more important in the moment, as well as the interactions of the people involved. In the end, the choice is instrumental—which identity is more useful, more powerful, in this particular configuration.

The intercultural "windows" phenomenon, in which people sometimes emphasize cultural differences and at other times submerge them, is well established in the anthropological literature. The Navajo, the Diné people of the American Southwest, illustrate how flexible cultural choices are made. Speakers of a larger language family that embraces sections of the Pacific Northwest and Canada, the Diné are relative newcomers to Arizona and New Mexico. In the Southwest, they "borrowed" many cultural elements from the Western Puebloans, such as the Hopi. Among the cultural artifacts so chosen was the practice of weaving. While the Diné may well have brought with them a tradition of weaving cedar-bark cloth, weaving cotton on an upright loom is clearly a Puebloan practice, done by the men of the villages. This practice was adopted by the Diné, and transformed again when sheep-herding was "borrowed" from the Spanish. Diné men became sheepherders, and the result of these multiple adaptations was that the weaving of wool blankets by women became a defining feature of Navajo life.

Further, the economics of weaving shifted with historical circumstances. Navajo blankets were prize trade items among Mexicans, Utes, and Plains Indians. This trade flourished in the classic period between 1800 and 1865, until several events once more transformed the craft of Navajo weaving: Machine-produced Pendleton blankets flooded the market, and Navajo territory was vastly constricted by U.S. military action against them. Traders came onto the land, acting as culture brokers between the women weavers and the Euro-American market. The traders bought the pieces, and also sold dyes, selected patterns, and encouraged new products. From 1865 to 1890 the traditional blanket began to fade from the market, to be replaced by more salable rugs. During the period from 1890 to 1920, rugs dominated production, although traditional garments were still woven for local use.

The evolution of rug design reflected complex cultural choices, sometimes minimizing Navajo identity and at other times making that identity the chief commodity. When Navajo identity was devalued, traders demanded reproductions of the Central Asian rug patterns desired by Victorian America. The trader Moore reputedly brought in tiles with Persian designs for his weavers to copy, to fit current consumer tastes. Navajo identity was irrelevant.

As attitudes toward Native Americans changed again, hand-woven rugs became fashionable once more. Consumers wanted rugs that were identifiably Navajo. The symbols that the Diné had traditionally used, drawn from stories of Spider Woman, Monster Slayer, and Born-for-Water, were restored. Pictorial designs with landscapes, range cattle, and feathers met the cultural expectations of the new consumer. Distinctive regional styles developed in which the Asian motifs demanded by earlier traders were incorporated into uniquely Diné designs. By the 1920s, traders fostered a revival of "authentic" vegetal dyes, rather than the bright, imported synthetics they had formerly promoted. "Navajoness" was the true commodity, rather than the rugs themselves (see Bassman 1997; Bennett and Bighorse 1997; Dedera 1975; Harmsen 1985; Wilkins and Leonard 1990; Pendleton 1975; Pendleton [personal communication]; Rodee 1987). Of course, to the weavers, all of these pieces were Diné, but the visible markers of their identity waned and waxed as required by the changing clientele.

The people of Silicon Valley also draw on various identities, at times maximizing the differences between them, and at other times minimizing them. A number of strategies for this cultural manipulation are evident and often are used simultaneously. Differences are maximized when inequality and resistance are greatest. Acts of discrimination force the categories underpinning the prejudice to be made explicit, and thus accentuate the plurality of the community. People negotiate these differences by creating corridors of sameness, in which contact with difference is minimized. People also isolate and channel these differences, trying to impose some sense of social order on the diversity around them. They create areas of life and institutions in which contact is restricted with members of particular groups. People also switch identities in midstream, minimizing one identity and exaggerating another. Differences can be masked when work—and particularly, technology—is made paramount. People can also create an alternative framework to make the individual the "unit of difference" rather than the group, placing potential social differences in the realm of psychology.

Using jokes, politeness, and a sizable dose of tolerance, Silicon Valley people have created an ethos of civility, where cultural gaffes are expected and forgiven. This ethos is justified by a very real instrumentality. One never knows who might be needed for the next favor, the next job, or the next son-in-law. Cultural borrowing is ever-present; the social presence of Asian, Latino, and numerous other cultures creates a repertoire of artifacts that become the property of all Northern Californians. Cultural gulfs can be narrowed by acting more like the "other." While these strategies make life easier on the individual level, they also pose deep contradictions and challenges to the creation of coherent communities.

Maximizing Difference

Not all of the tools used for managing difference in Silicon Valley are new social inventions. Maximizing social differences is a historically proven strategy for social control. Drawing a firm boundary between self and other works well for those with the power to maintain their position. Although the rhetoric of celebrating diversity is popular in California, people hold contradictory beliefs and behaviors. Interviews with Silicon Valley engineers and educators suggest that they are less than sure of the benefits of the changing complexion of the region. There is a great deal of ambivalence about cultural diversity, and this ambivalence is played out in both gross and subtle discriminatory actions. Racism and culturism, traditional mechanisms for controlling social relations, have not disappeared from the landscape.

It is useful to separate workplace discrimination by class. In her study of immigrant production workers, sociologist Karen Hossfeld examined the reasoning processes of managers as they spoke about their lower-echelon minority workers. Asian immigrant women were the preferred laborers. Filipino women were "meticulous and dependable"; Vietnamese ranked slightly below them, because of culture shock and language difficulties (1988: 278). These attitudes toward Asians contrasted with attitudes toward African-American applicants, who were seen as not "good enough" workers (ibid.: 271), and Hispanics, who were seen as lacking ambition (ibid.: 282).

Immigrant workers were believed to work for lower pay at "worse" jobs willingly, since they were "more desperate." Company managers felt that they were doing immigrants a favor by providing work (ibid.: 269). Hossfeld found that anyone who spoke with an accent was treated as an immigrant—regardless of birthplace—and placed in production work. Immigrants themselves internalized these attitudes to the extent that they believed that immigrants must "pay their dues" before "making it" in America (ibid.: 274).

This level of discrimination is augmented by more subtle cultural ambiguities. Discrimination reflects biased appraisals by organizations based on received gender and racial categories. Among middle-class professional workers, the stereotype of compliant Asian "model minority" workers creates a "techno-coolie" track (McLeod 1986: 114). New Asian immigrants are recruited into local firms through educational vectors, such as Stanford. Taiwanese are favored for their engineering expertise, while immigrants from Hong Kong are preferred as entrepreneurs. LEAP, an Asian American public policy institute, notes that a quarter of the total Silicon Valley workforce is Asian, as is half of the manufacturing workforce, and that technical skills and entrepreneurial connections are fueling the current boom (Park 1996). When asked about their success, Chinese immigrants themselves said "that it is their feeling that the Chinese work somewhat harder, or longer, or with more diligence, or with more company loyalty, than the average American employee" (McLeod 1986: 218).

Technical education, language barriers, and cultural styles tend to channel Chinese into professional, rather than managerial, positions. Six percent of the technical workforce in Silicon Valley is Chinese, but in management they represent only 4 percent (Saxenian 1999: 17). At Intel, 20 percent of the engineers are Chinese immigrants, and 80 percent of Cadence Design Systems' technical staff (Erasmus 1996: 180). In Hong Kong these Silicon Valley refugees called themselves "techno-coolies"— professionals channeled into technical work and steered away from managerial ladders. This restriction in career mobility motivates some of the "astronauts" who return to Taiwan and contribute to entrepreneurial activity in the Taipei-Hsinchu corridor, Taiwan's "silicon place" (Jung 1996b: 1E, 6E; Liu 1991).

Other forms of discrimination are less easily resolved. George, an African-American worker, notes that the "restriction in career mobility" is "the hardest part that a minority has to deal with. You keep trying and trying and trying, and you keep getting rejected, rejected, rejected. Not given the chance. That's how I feel about the racism." He goes on to add that it is particularly problematic since it is so covert, adding: "I think right now, with the era where its real subtle, like sexual harassment. Anyone who's conscious and knows all the outcomes of sexual harassment may still do it, but they'll be real, real subtle with it. They'll be so subtle that it'll be damn near impossible to prove." He goes on to predict: "That's how the racism is going down. It still exists, and I think it'll be in the Valley for a long time."

Beyond the familiar American-born types of discrimination, there are other competing forms of racism and culturism. Racism can be culturally constructed in many different forms, reflecting non-European bigotries expressed by and against other groups. Europeans and their descendants do not have a monopoly on ethnocentrism. The cultural assumptions of a highly diverse ethnic landscape reflect attributions and prejudices that overshadow the usual two-dimensional, black-and-white models of mainstream-other interaction. The conflict inherent in the interaction is made more complex by the "otherness" of all involved. Serbs and Croats squabble in Sunnyvale as well as in the Balkans. Twenty-five hundred years of hostilities between Vietnamese and Chinese still lurk underneath the surface at a research lab in Palo Alto. Jews look askance at Germans. Consider the following stories.

Lance is a first-generation Taiwanese-American who has been living in the United States for more than ten years. He was brought over by his father and stayed to complete his education after his father returned to

Taiwan. While a student, he started dating a second-generation Korean-American, Sue. When Sue's parents found out that she was involved with someone, they immediately flew to California from Chicago. To Lance's dismay, Sue introduced him as "just a friend" and assured them she had no intention of dating or marrying outside her "race." In front of Lance, Sue's parents talked about several eligible men that were waiting to meet Sue in Chicago. Sue and Lance avoided each other for some time and then decided to live together, but Sue insisted that they have separate phone numbers. Lance was forbidden to answer Sue's phone in case her parents should be calling. The apartment was kept in such a way that Lance's presence was not apparent. When Sue's sister visited, Lance had to stay with friends elsewhere. Lance was perplexed. He had not expected to encounter intra-Asian bigotry.

Note the attitudes of these Japanese restaurateurs. Both Japanese-Americans express anthropological folk models—culturally patterned stereotypes of other people. Kouzou has strong opinions—"I work with Mexicans . . . Mexicans are lazy!" He continues:

For example, every ethnic group tips differently. Japanese don't tip at all. Filipinos or people from somewhere around there are cheap so that they rarely tip. Mexicans don't tip, either. Only whites do courteously. Asians are bad, too. Japanese are bad, too. . . . They visibly look down on waitresses. . . . They kind of humiliate them. . . . Koreans aren't good, either. . . . If this is an American restaurant or something, they would be so "nice" to waitresses. Among Asians, they always try to be arrogant. If they go in front of whites, they would shrink. That's a bad habit of Asians.

Kenji gently echoes some attitudes, and disputes others, saying:

Since I have been working with a number of Mexicans, I always admire the way they are. I mean that they are so pure. . . . I feel so good to have a person like Jose as an employee. He has never spared himself working at the restaurant. For example, young Japanese kids only think about rewards for their work. Always a "If I do this, what would you give me?" kind of mentality. But [Mexicans] are different. I think that their living standard might be low materialistically speaking. But their living standard is very high in terms of their soul or spirit. . . . I guess Japanese are becoming lazier and lazier nowadays since we live in an affluent society. . . . Mexicans are too happy sometimes. They can be seen as living from hand to mouth. . . . That is how they are in a way, but it doesn't mean they are lazy or anything. I sometimes envy them. I wish I could live like them.

Jokes and jibes such as these about "lazy Mexicans" are traditional mechanisms for conveying superiority. Stereotypes, even those meant to be flattering, maximize and reinforce the differences between categories of people.

Given the prevalence of strong patrilineal cultures among the new immigrants, where descent through the male line is emphasized, gender status in particular becomes a focus for disputes. Should all aspects of one's cultural heritage be preserved, even gender discrimination? Indian immigrants are sending daughters to boarding school in India, hoping to elude the lure of American female gender roles and preserve the custom of arranged marriages (Mangaliman 2000). Young Vietnamese women report a conflict between wanting to retain some elements of their birth culture, but also wanting to reject the legacy of male dominance and control (Freeman 1996: 97–98). Rosalie, a mechanical engineer, wonders if she ought to have been offended when a Moslem manager approached her husband, instead of her, about her request for a new position. They had worked together for years, but in the end, he could not interact with her in a nontraditional fashion. Gender role conflicts enter families, workplaces, and consumer spaces.

People can emphasize their own cultural differences as a tactic for managing social situations. Note the attribution to culture of what might be interpersonal issues in the following story. Vanessa had married Dominic, a Mexican-American professional, five years before. Her teenaged children from a previous marriage came to live with them. Vanessa looked forward to a stable life as her own career began to flourish. Prior to the boys' return all the couple's social contacts had been with Dominic's Mexican-American family and acquaintances. Dominic is a socialist, and he disapproves of such decadent American frivolities as vacations, travel, and leisure unless there is a socialist political agenda attached. The teenage boys are self-sufficient and work hard on their household chores, but Dominic is never satisfied and sets rules concerning their clothing, bedrooms, bathrooms, and kitchen use. Vanessa and her sons are never consulted, but they willingly comply. Dominic continues to complain and begins to use "the silent treatment" on them. Although Vanessa's relationship with her sons improves, Dominic is critical of her child-rearing and disagrees with any attempt she makes to ease the tension, saying, "It's a cultural thing, and you would not understand," and accusing her of racism. Finally Dominic begins to demand rent from the sons, even though he had once promised he would never do so. Vanessa begins to pay rent for them, but after a couple of months

Dominic demands more per month. Vanessa is willing to do so for harmony's sake, but that harmony does not come.

Dominic's cultural difference may or may not have been the issue, but by citing his culture as the basis of the conflict, Dominic is underscoring their difference. Had she seen his behavior as personal, rather than cultural, she might have been less accepting of it. If his behavior is culturally based, however, she feels she has no choice but to submit.

Aaron, an ethnically Jewish engineer, uses his culture to shift the framework for group interaction dramatically.

We were six weeks away from beta. The date had not changed. We were exhausted. I was very short tempered, it was just bad. . . . We had a meeting starting at 5:30 on Friday; [our boss] looked at us and he says, "Okay, six weeks away from beta, you guys look tired. You're exhausted aren't you?" "Uh huh," [we said], shaking our heads, just looking miserable. . . . [Our boss] asked, "Do you guys need a rest? You need to take a break, guys? . . . Tell you what, I want everybody to go home and take your wife, significant other, whoever, out for dinner and a movie on [the company]. We'll pay for it." The first thing that went through my mind, of course, is that, baby-sitter starting at 5:30 . . . okay, we'll borrow that time. [The boss] adds, laughing, "Go home and just relax, get a lot of stuff out of your system. And I'll see you guys at 9:00 tomorrow morning." And the silence was this wall. . . . No one's saying anything. I looked up and said, "I cannot work tomorrow. It's against my beliefs. I can't work tomorrow, it's my Shabbat." And the turnaround, the snapping of heads all around the room, was mixed with the boss who backpedaling as fast as he could and the other people in the room are going, "Whoa, can I join?" It was just cool. . . . [I said], "I can work Sunday, but I cannot work tonight and I cannot work tomorrow." [The boss says], "Oh no, I didn't mean it. Okay, everybody comes in on Sunday." And he had bagels with lox with cheese.

This interplay illustrates the power of invoking difference. The tactic works because mistreating people on the basis of difference is a cultural faux pas. In such a climate, manipulating identity-based differences is a reasonable tool.

Channeling the Flows

Creating corridors of sameness is another tactic for managing the identity diversity. Ethnic enclaving is a well-established American tradition, one that still can be found in the Valley. "Japantown," "Little Taipei," and "Little Saigon" all reflect a geographic concentration of difference. Jennifer, a high-tech trainer, notes, "But just down a couple of blocks away is where there's a more Hispanic neighborhood. There's a whole strip. They kind of keep things separate. A whole strip of Spanish butchers, bakers, whatever catering to that more Hispanic taste." Japanese *chuzaiin* keep their social life largely within Japanese circles; German immigrants can find the foods from home they miss. However, the neighborhoods are not so much physically segregated as emotionally negotiated. A monocultural comfort zone is established in spite of conflicting values. The social barriers may be erected from many directions—by the outside community, from within the ethnic community, or by the individuals themselves. Avery, a young non-Jewish man with a degree in child development, taught in a Jewish preschool. After several faux pas regarding kosher food preparation, he was denied promotion. While he was qualified in his work skills, the principal noted that the children "needed a Jew in the classroom to guide [them]"—to ensure their cultural competence. The boundaries are maintained.

The differences to be contained do not have to be ethnic. People seek out communities of people who are like them in age, gender, hobby, class, and religion. Jeff, a software engineer, notes, "I tend to think of the Valley, most of the time, as a place where everybody's an engineer, and of course that's patently false and foolish. But I move in the engineering circles, so those are the people I know and those are the people I mostly notice." The lives people live, working with people of similar education, interests, and class lead to a perception of sameness. When asked if the region was diverse, some people said it was not, citing these corridors of homogeneity as proof that the region's diversity was merely demographic, not experienced in daily life. Jennifer, an instructional designer and trainer, comments on the culture shock she and her colleagues experienced at a corporate outing at the Great America theme park: "Wow! We're really in an ivory tower when we're at work . . . not only culturally but monetarily. . . . And to actually see lower classes! And I take the train to work. . . . You see homeless people camping out. And you see run-down homes and trailer parks and so there is another, a whole other world out there. But it's easy to isolate yourself from it."

If one cannot simply separate oneself from people who are different, then another tried-and-true mechanism for managing complexity is to channel people into particular segments of society. Castes create specific economic and social niches for whole categories of people. There are echoes of this strategy in Silicon Valley. Cambodians work in donut, and, increasingly, in bagel shops. Sikhs run nearly all the taxi businesses (Nhu 1997: 1B, 2B). If educated Chinese immigrants are barred from the existing "old boys network," they can become entrepreneurs, particularly at the early stages of funding, or become small subcontractors (Park 1996: 164, 172; Saxenian 1999: 20, 57). Karen, a research and development project manager, observed that she encounters Latinos in the organization and beyond, but only in certain niches. She notes: "It's not hard to notice [that] the nannies and the gardeners and the maintenance people are all speaking Spanish."

Latino immigrants came to Santa Clara County for its agricultural and cannery jobs. A second wave of Mexican immigrants came in the 1960s and 1970s for the unskilled jobs of the electronics industry. Then a third wave came to staff the service-related jobs. At the beginning of the 1990s, Latinos were 80 percent of the clerical and operating jobs in the low-wage service sector (Zloniski 1994: 2310). Fifty-four percent of the janitors of Santa Clara County were Hispanic surnamed (U.S. Census Bureau n.d.).

However, there is an inherent difficulty with channeling differences by linking ethnic groups with economic niches. There are too many competing criteria for identity for one category to dominate interactions. Ethnicity, education, gender, or social role compete for the foreground in any particular interaction. Once the barriers around the economic niches are broken, people can then emphasize or de-emphasize other commonalities or differences. I was told of an incident that highlights this phenomenon. In a large high-tech company, two Latino engineers joke with each other in the corridor before a meeting, commenting on the rarity of Latino engineers and feeling a spark of camaraderie. Yet during the meeting, they argue vigorously. Afterward they ask each other, "What happened in there?" and finally realize that for that specific time and place, the fact that they are different kinds of engineers divides them more than their common ethnicity unites them.

Minimizing Differences

Sometimes it is not pragmatic to emphasize cultural differences; the social costs are too high. I have observed two strategies for minimizing cultural differences among Silicon Valley's people. In the first, potentially troublesome distinctions are pushed to the background as alternative frameworks are created to override them. These constructs may be based on work, professional identity, organizational culture, or techno-enthusiasm; and a host of psychosocial tools form an alternative "window" in which interaction takes place. A second strategy fosters exposure to other cultures, cultivating an etiquette of civility in the face of cultural difference.

The penetration of work into all of life has some significant consequences for managing culturally based differences. Work creates a new social framework in which the differences can be put in the background. It enhances an ethic of expediency and instrumentality in which getting the job done is more important than any other social activity—such as preserving ethnic identity. The workplace reinforces an alternative language of difference, in which distinction is individualized—attributed to personality, not culture. Workplace etiquette—"acting professionally"—supports an ethos of civility in which it is not appropriate to verbalize prejudices. The workplace also creates a social setting in which people's cultural repertoire is enriched, allowing them to borrow new cultural information to add to their mental tool kits, adding new artifacts to their daily lives. Jaime, a Latino engineer, summarizes a prevailing sentiment:

Silicon Valley attracts everyone. I think the commonality there is technology. . . . It seems like a lot of the people that you kind of come in contact with share the same belief of technology . . . [breaking down] a lot of the culture barriers or racial barriers . . . more or less on a day-to-day basis of coworkers or just standing somewhere in Fry's. Regardless of your color or race, . . . pretty soon you get so caught up with talking about whatever it is you're talking about and actually stop realizing what color the other person is. . . . For me that's . . . what attracts me about Silicon Valley.

Jeff, the software engineer, whose observations we have already encountered, notes:

There are a lot of Asians on my floor at [the company]. Perhaps every other door is Asian or Indian. And a smattering of others. You can find people from the Middle East, people from Europe, and blacks. . . . The difference in culture doesn't show up so much at work, typically. It's just a matter of trying to understand someone through his accent, and some people have them and some don't. But you're all talking computers all day long, and that's the culture. Other than that, it's just a matter of, "My God, what is he eating for lunch?" That's about the only thing that intrudes, or the decorations that they have on the walls of their cubes or offices. The business kind of swamps the cultural stuff unless you become friends and start seeing each other outside. But you do start becoming aware of things, you know. Not only is it Christmas, it's Hanukkah. And, "Oh, it's Chinese New Year now—he won't be in to work," or "He's celebrating." You just start to pick that stuff up. It's kind of nice. But at work, mostly people talk work.

His observations echo those of an Asian-American journalist who notes, "When we're together . . . say for a meeting, you're so focused on a task that you're not sitting there saying, 'Oh wow, there's a [cultural] difference here.'" The primacy of work minimizes the perception of difference.

Work also provides a platform for eroding traditional differences. For example, Clare is sensitized to difference by her lesbian identity and the African-American ethnicity of her partner. She is conscious that her educational and professional work with women in technology is bound up with her own political efforts to change the status of women. She says, "It's much more dramatic than if you're trying to make a personal change outside work. . . . If you can tie the two together, then you do get . . . the ability to change."

People manage cultural differences by creating new ways to interpret them, explaining away behaviors once understood to be cultural as a function of individual choice. By attributing differences to individual personality, rather than to culture, people can learn alternative stereotypes and minimize old prejudices. Deep discussions of politics, religion, and social difference are avoided. Instead, people emphasize individual and psychological nuances. They customize their environments to create "sanity place[s]" that reflect their individuality. People "want their individuality to be respected."

The Myers-Briggs Type Indicator, or "MBTI," personality inventory was the most frequently encountered alternative framework. This indicator is widely used in corporate training—as one human resource director said, "Every sizable corporation with a human resources department can or does use the Myers-Briggs." The inventory creates a typology of sixteen different personality types, based on four dimensions: (E) extroversion or (I) introversion, (S) sensing or (N) intuition, (T) thinking or (F) feeling, and (J) judging or (P) perceiving (Eng 1997: 6G; Thorne and Gough 1991). For example, an "INTJ" represents the typical engineer. She is deliberate, logical, serious, individualistic, indifferent to differences, in control of her own impulses, proud of her own objectivity, and tends to delay gratification, perhaps unnecessarily (1991: 88).

People described the benefits of interpersonal training courses that allowed them to understand themselves and the processes of change around them. They used these tools to understand themselves and their relationships, at work, at home, and in the community, and to find potential partners and mediate spousal conflicts. The language of psychology has reshaped the corporate order to emphasize individuality (Pfister 1997: 200).

One multinational high-tech organization instituted a program to foster a set of "corporate values." The program was developed by the upper-level human resources personnel to provide nonconfrontational tools for social change, and the values were to be implemented by everyone from the CEO down. The program was named "CREATIVE," an acronym for Customer focus, Respect, Excellence, Accountability, Teamwork, Integrity, Very open communication, and Enjoying our work. Packets were made for employees describing each value and what it meant to the organization. A pen was designed that would display a different value each time it was clicked. Coaches and counselors were ready to help an angry communicator shift his values. Work teams were encouraged to function more amiably, and confrontation was equated with disrespect.

Everyone was enthusiastic, except the people who said it didn't work. These people worry about the apparent inconsistencies in implementation, and feel that upper managers exempted themselves from the values, particularly the "very open communication." The critical point here is not whether or not the program was effective, but that it created an artificial set of alternative "interpersonal" skills by which people were to operate—a set not based on obvious ethnic, cultural, or gender-based criteria. Instead, the program employed an operational style that is becoming part of the folk anthropology of the Valley, replacing an aggressive "Western" communications style with an ethos of cooperation to produce corporate harmony, a style imagined to be "Asian."

In Silicon Valley civic discourse links diversity to increased productivity and treats it as a reason to celebrate. Joint Venture's blueprint for the region's future refers to the "engine of diversity," declaring that "our inclusive society connects people to opportunities" and that "Silicon Valley's greatest asset is the diversity and spirit of its people" (Joint Venture: Silicon Valley Network 1998: 30–31). The phrase "engine of diversity" is repeatedly used in public rhetoric. The context in which it is used makes it clear that it refers to two types of diversity—ethnic and economic. Both are celebrated as the drivers of innovation.

Large-scale public displays of diversity, as opposed to the private experiences discussed in the last chapter, center around "celebration." While our interviewees recognized these large-scale events as "superficial" and "trite," they also viewed the public display of music, dance, food, and arts as an opportunity to "participate in and observe" communities that might not otherwise be accessible. Events such as the Chinese Summer Festival, Mountain View's Afribbean Festival, the Japanese Obon Festival, the Vietnamese Tet Festival, and the Dia de Portugal Festival allow various communities to celebrate themselves as well. The target audience is not only the greater community but also the next generation.

Two events are explicitly multicultural, devised by the larger polity to create awareness of diversity. One event is the annual Fourth of July San Jose America Festival. It includes a cornucopia of multicultural symbols—African drumming, Navajo fry bread, Southeast Asian needlework. A visiting mother opines: "I want [my children] to learn acceptance of what's different, understand what's different, not be afraid of what's different" (Rae-Dupree 1995: 1B, 4B).

In 1998, Santa Clara County organized its first "Unity in Diversity Day." While the requisite tasty snacks and multicultural arts performances—among them the Ballet Folklorico of Mexico and Filipino Folk Dancers—were included, the agenda was more explicit in promoting civic unity. Unity in Diversity awards were presented by the board of supervisors. Flyers recruited volunteers for the Billy DeFrank Lesbian and Gay Community Center. The Anti-Defamation League distributed articles about anti-Semitism. Contributions were solicited for medicine for Iraq, and information booths told German-Americans about their history and affiliated organizations. The Hispanic Chamber of Commerce distributed invitations to their Directory Celebration Party and Networking Forum. At the community level, these events are orchestrated to raise the visibility of the "engine of diversity," and to promote interaction "to all who would partake" (Tran 1998a: 5B).

Individual ethnic communities, of course, have their own celebrations. In one division of a high-tech company, Huy, a Vietnamese test operator, talks about how he and his Vietnamese coworkers get together informally to celebrate Tet, the Vietnamese New Year. He says: "[We] talk together. We make a decision [about] what time. What day. We celebrate. We take one hour, two hours on the roof, somewhere in the garden, outside the picnic area. We bring . . . some special food for Tet. We eat together. . . . We talk about the Tet we remember, how the Tet was in

Vietnam many years ago." He then goes on to say that the celebration is not just for Vietnamese, but includes the Chinese workers, since they share the "same" culture, and it is open to others as well. Vietnamese rice dishes are offered in office lunchrooms, next to french fries.

On the other hand, Huy's company offers a more organized, sponsored celebration of Cinco de Mayo. At lunch, the company provides Mexican dance music and serves Mexican food in the cafeteria. The Tet celebration lacks formal sponsorship, and Huy suspects that this is because there are fewer Asians—that is, Vietnamese—in upper management to promote it. Nonetheless, whether by informal association or formal observance, workers are exposed to many cultures through these events.

Gathering into a homogeneous group in public is considered potentially impolite. Mary Ann, a consultant, observes: "The Indian contractors work together and lunch together and don't socialize much with others. It makes me feel restricted—I don't always feel like I can jump in and be friendly when the group is too predominately one race." Mary Ann reacts to this by trying "to find reasons to begin to talk and become a part of the group."

This story is consistent with reports of Silicon Valley teenagers who say that they feel troubled by a monocultural or monochrome grouping, and view a multicultural group as the comfortable norm. Erin, a Euro-American mother, took her young child to Chinatown in San Francisco. He "started feeling really uncomfortable, and he said, 'What is this, mom?' And I said, 'This is Chinatown. Chinese people live here.' He said, 'They're talking a really funny language.' And he saw all the food hanging from the wall, and there was different looking food, and he said, 'Mom, let's get out of here.'" What is noteworthy is not that the child had the initial ethnocentric reaction, but that Erin took him there in the first place and believed it was important to teach him to react differently.

Creating an atmosphere of civility can mask potentially damaging prejudices. Humor, somewhat problematic in cross-cultural situations, is used to create relationships. Kathleen, a marketing manager from Ireland, tells how she used humor to handle a difficult management situation. When someone in one of her meetings makes an inappropriate or offensive "smart remark," she tosses that person a stuffed bunny. That person must keep the bunny until someone else "earns" it. This "lightens the atmosphere . . . but puts a context on it at the same time." David, who works in a widely diverse multicultural setting, employs joking relationships strategically to form working relationships. He uses teasing, horse-play, and quips, aiming each form of humor at the appropriate audience. His horseplay is reserved for fellow African-American men; witticisms are used with others, in more restrained situations such as staff meetings.

The ethos of civility requires an intolerance for intolerance. Jennifer, who grew up locally, talks about how she was raised to disdain racism, but her husband, a migrant from the Midwest, violates that norm. She comments, laughing, that "sometimes some of the things that come out of his mouth are not acceptable to me . . . but he's getting California eyes, too." A migrant from Florida notes that only when she started working at a public television station did she begin to learn about the gay com-munity. Now the prejudices of her peers and family in Florida seem "alien," and since she doesn't "have the energy to invest" to defend her tolerance, she simply sees less of them. Kristin, a teacher at a multicul-tural elementary school, states her attitude succinctly: "I have so many languages in here . . . probably ten different languages. Languages that I hadn't heard of until I started teaching this class. One is Singhalese—they speak it in Sri Lanka. One is from a certain part of Ethiopia. . . . It's really interesting, and it adds to who I am, what I know, and what I can appre-ciate. It's also made me . . . frustrated or negative towards people who don't celebrate the same things that I celebrate."

Toni, a marketing manager, repeats the idea that working with differ-ent cultures has caused her to see things in a different light and to culti-vate a new sensitivity:

Where I worked . . . there were a lot of Japanese people. . . . This one gal would always bow to me passing in the hall, and I always felt funny saying "Hi" to people in Japanese, since *Hai* is not "Hi" there [it means "yes"]. . . . There was an Indian guy who worked at [the company], and he always saw me making my tea from the tap, the hot tap water, and twice he told me I shouldn't use that wa-ter because it wasn't as purified as some of the other sources of hot water, and I thought that was interesting because that was definitely something you don't worry about here. . . . Actually it strikes me, that at [the company], there were all these, mostly Asian, but Indian, Iranian, other people, and how they fit in so well.

Another person comments that she is "intrigued" with Asian cultures that she has encountered in the workplace, with "their language, . . . polite mannerisms." She adds, "I've even had people that I have interviewed that will bow to me—it seems so nice. I don't know, it's different."

The large number of Asians and the visible diversity of the region's population allow emigrants from Asia to feel more comfortable with themselves. Sima, the wife of an Indian software engineer, tells the anthropologist how, unlike her experience when she lived on the East Coast, "there's so much of diversity here I feel very comfortable wearing Indian clothes and walking on the street." Another Indian woman notes that the only comment she gets when she is wearing her sari is that her fabric is "pretty." Indians contrast the prejudice they may have experienced abroad with the attitude in Silicon Valley. A South Asian entrepreneur cites the "strong racial prejudice against Indians in London, compared with the much more open attitudes she found when she first came to California" (Janah 1999: 1E, 7E). Indian cultural accouterments have spread to the rest of the community, as non-Indian junior high school girls decorate their hands with henna.

Cultural differences add to the artifacts and ideas from which Californians can draw. A worker at a nonprofit agency reflects on his various identities. He was raised in a conventional Euro-American, middle-class milieu. Now he actively participates in Chicano groups, immersing himself in that set of cultures as well as identifying with Christians, and with other gay men. He notes, "I don't really listen to white music or eat white food or that . . . I prefer to take on other cultures with respect to where I live and my gay identity."

Lily, an interior designer, notes that "Latin-Americans and Asian-Americans will . . . have artifacts which represent their culture . . . spirit gods and the shrines, you know, just the little images." The same images, however, can be found in the homes of Chinese assembly workers, and in the offices of non-Asian engineers. The penetration of Asian cultural elements allows Gus, a corporate consultant, to matter-of-factly refer to himself as a "typical Okie Buddhist." Although not an Indian, Clare talks about the statue of Ganesha [a Hindu god] in her office, saying, "You pray to Ganesha . . . the remover of obstacles . . . [who also protects] your journey and spirit." For inspiration, Tom draws on a mixture of Hopi kachinas and a Japanese Zen mini-garden in his office shrine. He also hopes that his daughter will marry an Asian who will truly understand transnational extended families.

Cultural Instrumentality

People draw on the many cultures they see around them instrumentally, taking artifacts from one society and food from another, "Asianizing" and "Hispanicizing" themselves as needed. These choices are not random, but reflect an instrumentality that recognizes that the greatest capital in Silicon Valley is human. Silicon Valley values "a work culture . . . [that] insures that what you do adds value to the organization." People work on their own cultural competencies, their abilities to work with people from diverse cultures, as part of that value adding process. Jan, a Dutch sojourner, discovers that he "would actually tailor [his] questions and behaviors for the people that are in the room." He tells us that he switches his communication style when questioning Taiwanese or Indian colleagues, becoming less or more aggressive in his tone. "I've learned that lesson. I've definitely learned that part." Another person worries about having "enough diversity in my network." Stan deliberately recruits new students from the University of Texas in El Paso, Texas, where there are many Latino students, not only because of the technical reputation of the school but also because he can find job candidates who can add to the diversity of his group and his software company. Irene, a diversity consultant at a major computer company, explains that they need to create an atmosphere in which every person feels respected—respect being defined differently in diverse cultures—in order to maximize workers' contributions to the company. She explains how she convinces people to consider diversity important, promoting the notion that diverse workers create links to different national and international markets (see also Casey 1995: 101). The organizations need cultural agents—"liaisons between different cultures"—that can broker interactions with Latin Americans, Chinese, Koreans, and Southeast Asians.

Mark is a manager of information systems at an international company that designs and manufactures computers. He states the case for instrumentality succinctly:

Dealing with a cross-cultural organization . . . [it] would be interesting to figure out whether your overall efficiency is higher or lower. I know that you have to put more effort and resources into the day-to-day operations. . . . Say your daily operations [are] $100.00 . . . that's just what it takes everybody to get the organization going, and say you've got to add an extra 12 percent for additional time spent for discussion . . . for the overhead involved with having to be aware of cultural issues within the groups. Now you're at $112.00. But another culture has different ideas or comments or they're more sensitive to a certain portion of a market than someone else would be, so they add value back in, so they take N percent and contribute that back into gains. I don't know how that math works out between how much additional cost you have dealing with multicultural [workers] versus what is the value that is gained by being multicultural.

The cultural instrumentality in Silicon Valley minimizes the costs involved in managing cultural differences. I told this story to a researcher from Northern Europe who noted insightfully that the cost might be considerably more than $12 outside Silicon Valley, where organizations may lack the cultural tools or the motivation to place such differences in the background. Silicon Valley cannot afford the luxury of losing that marginal advantage.

Carl once had a major cultural conflict with two Indian colleagues. They persisted in treating him hierarchically, as their "servant," and Carl found that situation "unsatisfactory" and left for another job. However, in the next company his new boss and counterpart were again Indian. His job involves localizing networking systems to India. He cannot risk harboring ethnocentric notions. As he so eloquently puts it, "Dealing with the variety of people, all the cultural issues, it is impossible to know enough of the cultural issues to not step on toes. So the solution seems to be for people to get used to wearing steel-toed shoes. People are getting . . . more tolerant."

Two dimensions are critical for cultivating cultural competence. One is to become familiar with the particulars of specific cultures—largely by exposing oneself to different cultural contexts. The other dimension is more amorphous. Generic cultural competence is gained when people learn to accept that assumptions may not be shared. They learn to translate an understanding that core values differ into acceptable actions related to relationships, time, and so forth (Nolan 1999: 2–8). Jan, a Dutch engineer, discusses the strategies he uses to get the people he trains to gain the latter skill as they are being exposed to diverse cultural content:

We have this multiethnical, multicountry, multibackground, multiwhatever group here that has to find a common ground of working relationships, and that's a very interesting phenomenon. We're moving towards that, but at a very slow pace. . . . Different pacing, different reference frames. The very first thing that you do with those people . . . "Show them the mirror. . . ." Show them examples of how things can be different . . . and so you have them think, "So here's a different way of approaching it." What you hope to be able to tap [into is the] flexibility to see [the differences] as opportunities and not as threats. . . . "There's all sorts of things out there I don't know and I feel estranged." Or people can say, "Oh, that's new. I've never approached it that way. . . ." If you go through this change where you have a multicultural situation and you want to get it to high productivity, you have to basically broaden people's horizons, take their "eye claps" [his hands move like blinders off his eyes] off. Face them with different situations, different approaches, and then you'll see that most of the people . . . will see it as a positive experience. . . . There will be a certain percentage that gets offended by it and gets dysfunctional.

A wide range of efforts, both individual and organizational, are made to introduce culturally specific content to the community at large. Maha El-Genaidi founded the Islamic Networks Group to explain Islamic beliefs and life to schoolchildren, fielding questions about prayer, Ramadan, fasting, and her *hijab* (head scarf) (Hendricks 1997: 1E, 3E). Linda, a career woman and mother, also embodies this idea when she takes her "Mama Schwartz Traveling Jewish Show" into classrooms to expose children to different Jewish ceremonial cycles, including the celebration of Hanukkah.

The cycle of holiday celebrations is a deep feature of American life (Naylor 1998: 111) and is being altered by the inclusion of holidays from other cultures. Tom notes that in his organization they have an incredibly diverse group: "We've had to stop using European ethnic names for holidays. So we don't have 'Christmas'; we have the 'winter holiday season.'"

A framework of common experience is necessary to facilitate cultural competence. One way to accomplish this is to increase the number of personal interactions among people who do not ordinarily interact. Dierdre, who facilitates community work in a computer corporation, does this by recruiting Vietnamese and Latino leaders in the Girl Scouts.

Other efforts take a more open-ended approach, simultaneously sensitizing people to the unknowns inherent in an intercultural encounter and preparing them to face the inevitable culture shocks without taking offense. Irene, the diversity consultant, relates it to the cartoon insight that "we are all the same and all different." She talks to young people in this way about diversity, hoping to make them aware, without "beating [themselves] up," of their assumptions, biases, and prejudices.

At a higher level, Silicon Valley itself struggles to recognize its own identity as a culturally plural community. In order to incorporate both technological focus and identity diversity into an "efficient" and "effective" community, Silicon Valley is reinventing itself as an innovator in cultural competence. Glenda, a high-tech artist and graphic designer, ponders how she would photograph Silicon Valley "community." She imagines juxtaposing an image of a micro-chip—"all those tiny bits and pieces [of] information and technology"—with images of people. The people would have to be diverse, since her clients are "global." The imagined image would reflect the experience of "walking down the street and hearing four or five different languages all going on at once."

Silicon Valley's educators are in the forefront of creating community in the midst of diversity. They recognize that teaching in a multicultural classroom means changing the way courses are taught. Simply teaching in a different language will not solve the problem. According to the most recent figures—the 1998–99 English Learner (EL, formerly LEP, Limited-English-Proficient) students, and the 1997–98 Fluent-English-Proficient (FEP) students—46,581 Spanish-speaking students attended schools in Santa Clara County. Another 54,499 students spoke Vietnamese, Tagalog, Cantonese and Mandarin Chinese, Korean, Punjabi, Khmer (Cambodian), Japanese, Farsi, Ilocano, Portuguese, Chaozhou and Taiwanese dialects, Russian, Hindi, Hebrew, Arabic, Lao, Gujarati, Urdu, French, German, Samoan, Assyrian, Croatian, Polish, Italian, Dutch, Armenian, Greek, Rumanian, Tongan, Serbo-Croatian, Burmese, Cebuano, Indonesian, Turkish, Mien, Hungarian, Hmong, Serbian, Ukrainian, Pashto, Chamorro (from Guam), Kurdish, Albanian, Chaldean, Marshallese, Tigrinya, and Mixteco and other Native American languages (California Department of Education 2000). No one cultural group or language provides an alternative to English. "Sheltered English" environments are one way to accommodate the linguistic differences in the classroom. These classroom environments present information in a variety of ways, not only with verbal lectures but also with visual and tactile aids to facilitate communication. Collaborative techniques force the children to work across differences. Teachers and students learn that in a plural environment the potential for unintended offense is great, and no behavior can be assumed to be inoffensive. These children are being given an experiential crash course in cultural competency, as they become sensitized to the idea of difference. Students interviewed in general education courses at San Jose State University cite the interaction with culturally different peers as one of the greatest assets of that institution (Darrah 1997). Cultural competency becomes one of the tools needed to be effective in their future work teams, transnational interactions, and future civic life. Reminders of the need for this competency come in civic events, in public celebrations, and in corporate discourse.

Cultural competency is increasingly viewed as a vital component of the new workplace etiquette, as work organizations struggle to inculcate cultural tolerance in their employees. The deportment curriculum is designed to make a plural workplace "efficient and productive." The content of the class may vary with the audience. Diversity classes, designed to make workers conscious of the rights of protected classes (race, ethnicity, gender, disability), are aimed at the ordinary worker (Carrico 1996), while courses aimed at easing global intercultural interactions target the professional and managerial elite.

Organizations offer classes on how to interact with specific cultures, particularly that of Japan. Barbara, an executive admin, recalls the difficulty she had enrolling in a company-sponsored course on interacting with the Japanese. The class was reserved for engineers, but she persevered, arguing that she, too, had significant interaction with Japanese sojourners and needed to be sensitive to their customs. Such culture-specific courses, however, do not go to the heart of the dilemma of cultural plurality.

Learning the surface customs of one culture can mask ignorance of a wealth of hidden cultural differences. The Japanese are not the only "other," and simply learning, for example, not to show the sole of one's foot does not prepare Americans for the more complicated reality of Japanese social hierarchy. A useful code of civility must embrace the very ambiguity of intercultural experiences.

Luke, an experienced corporate instructional designer, articulates how he incorporates cultural competence into all his workshops. In his words, he looks for the "lowest common denominator." When he teaches, he considers who among his audience is most sensitive to classroom organization (group versus individual learning), reward structures (team praise versus individual accolades), and interaction with authority. He then structures learning around those participants who are most at a loss when encountering different cultural assumptions. In this organizational equivalent of "sheltered English," the most culturally vulnerable are able to learn, and the other students in the class also broaden their repertoire of interactional skills.

Technology designers need to be conscious of how technology is used in different cultures. For example, Michael's task was to adapt a printer to the technology practices of Japan. He knows that most of his Silicon Valley coworkers never turn their printers off. He also knows, however, that the Japanese do turn them off at night. This meant that the Japanese needed a quick power up and down time, a feature not needed for American printers. Michael needed to pay attention to both technology and to culturally distinct consumer practices to do his engineering work properly.

Roald, a Norwegian, is a localization engineering manager. His job is to make sure that technologies reflect "local" cultural understandings for particular parts of the world, so he needs to understand the technical implications of cultural differences. Consider the icon on screen displays that alerts people to the presence of electronic mail. In the United States, a symbolic mailbox conveys the information. Is this icon appropriate, however, for cultures in which real mail is not delivered to individual household boxes? Engineers must understand both the cultural and the technical aspects of a product to design commodities that their global customers will want to buy.

Just as global business thrives on the localization of products to select parts of the world, cultural localization offers new niches. A new multicultural mall has been proposed for East San Jose. Like the open-air markets of other parts of the world, Asian noodle shops and Latin American zapaterias will be placed side by side. Overseen by a Vietnamese developer and partially funded by the San Jose Redevelopment Agency, the new "mercados" are the "prototypes of the multiculturalism we see today and what's going to happen" (Ha 1999: 1E, 4E). These malls are centers for immigrant life (Tran 1998b: 1B, 5B). Such centers are hardly original; similar markets can be found in the archaeology of Chinese sections of Gold Rush–era towns. However, the conscious development of such markets as a form of product localization, and the view of them as "prototypes" for the construction of new community forms, is very typical of Silicon Valley.

Another example of this localization is the construction of a Chinese language office—offering services in both Mandarin and Cantonese—in heavily Asian Milpitas, by the investment firm Charles Schwab. Silicon Valley has the highest level of investment activity in the United States (Chen 1997b: 1C, 3C; Ha 2001: 1C). Schwab wants to tap into the $1.2 billion in assets held by local Chinese-Americans as well as make use of their global connections. Culture can be used instrumentally at the organization and community level, as well as by the lone individual.

At the same time that Silicon Valley leadership mobilizes resources to minimize and use cultural and social differences, there are still structural and psychological forces that seek to emphasize these differences. Cultural competency—feeling easy with differences, and being able to observe and adjust to new circumstances—is a powerful asset for Silicon Valley workers. It is not, however, without problems for those who would employ it.

Bringing Community Home

Exposure to other cultures in public venues—school, work, shopping, and public events—offers only superficial acquaintance. Pierre van den Berghe, in his classic work on institutionalized racism, makes a clear distinction between formal and informal interactions in daily life (1967). Formal, structural interactions are permitted in hierarchical, pluralistic societies in work, school, and public settings. These interactions are clearly defined, and avoid the informal, intimate domains of friendship and family. Dominant and subordinate groups "know their places." In this way, identity categories are clearly maintained. While this model can only partially be applied to California's cultural landscape, it does raise some interesting questions.

Dan's story illustrates the push and pull of formal civility and intimate discomfort. A longtime resident of the Stanford area, Dan describes the cultural differences in his neighborhood with contentment, noting his Japanese, Taiwanese, Indian, and Russian neighbors. He comments: "It's just interesting to see how they do things differently than we do." He notes: "When we bought the house, and we were just, had just been here for a few weeks, I had a really very vivid experience. A group of people came to the park to have a picnic. And this brought up all my territorial instincts. I was standing right here and those people were over there, and I was feeling like I was being invaded by these people." He did not, however, have these territorial feelings when he was not on his own property. Only when his intimate space was threatened was he uncomfortable with the chaos outside his corridor.

In Silicon Valley different conceptions of formality, propriety, and intimacy make the maintenance of clear identity boundaries impossible. Elliot talks about the difficulty of facing cultural differences when they intersect more personal issues. What happens when the Korean woman he supervises starts asking him what to do about her oppressive husband? Her husband takes the money she earns and gives it to his parents. When she complains, her minister supports her husband wholeheartedly. Elliot can tell her about California's community property laws, but "where personal stuff . . . gets beyond the work environment. That's where cultural things start showing up."

This quandary is especially significant because in Silicon Valley the boundaries between professional and personal are so permeable. Work and home are difficult to separate. As mentioned in Chapter 2, one of the strategies for building trust with coworkers—often strangers—is sharing personal information. Confidential communication may easily lead to discussions of culturally sensitive topics.

Teachers, machinists, engineers, and administrators socially intertwine work life and home life. This can lead to an "intimacy" in which cultural expectations may be violated, as the following story illustrates. Glenn had been friends with Eduardo, his Latino coworker, for almost two years. They worked together in the machine shop, drank beer, and ogled women. When Eduardo brings Glenn home to dinner, Glenn remarks that Eduardo's wife is an "absolute 10." After that evening, Eduardo angrily refuses to interact with him. What is an acceptable comment at work about strange women, is utterly shocking at home about one's wife. Glenn thinks he is complimenting Eduardo, but instead Eduardo feels insulted. As long as they keep their banter at a "surface intimacy" no norms are violated. It is only when they try to have a closer relationship that their companionship goes awry.

Yael, a female Israeli professor, had been in the United States for twelve years. She obtained her doctorate, married an American, and started a family. She became good friends with her coworker, Anabel. When Yael's mother became seriously ill, she went back to Israel to be with her. As she returned to the United States, her mother died. Yael decided that it would be too hard on her family in the United States to return immediately to Israel. She finds grieving hard, for she feels little support from her friends and colleagues, especially Anabel. Yael says, "I felt that no one was reaching out to me as they would in Israel." Anabel sends Yael a Hallmark card and grants her privacy, believing this offers comfort to her friend. Unknowingly she only accentuates her ignorance of what Yael considers appropriate intimate behavior—a "true friend" would have grieved with her openly and personally. Until the crisis, the culture of work had masked differences that would have undermined their friendship. Faced with the need to connect more intimately, Yael discovers that both grief and friendship are expressed differently in Silicon Valley and Israel.

Casey, a successful realtor, believes that if he could only convince the wives of prospective buyers to buy the luxury homes he sells, the sale would be "in the bag." His cultural experience is that once the wife is "sold," she can always convince her husband. Therefore he always seats the wife in the front seat of his car on the way to view a property. This strategy works, until he uses it with an Indian couple. They are furious with the realtor's display of inappropriate intimacy with the wife and his insensitivity toward the husband.

Social roles differ from culture to culture, and attempts to create "personal" connections with coworkers may backfire where cultural expectations clash. Pooja, an Indian intellectual turned beautician, is scandalized when her coworkers invite her to a birthday party that features a male stripper for entertainment. Mariana, a Filipino clerical worker, is perplexed when Vasili and Dimitri, her Russian immigrant employers, treat her harshly at the construction site where they work but warmly charm her in a social setting. Miguel, a Latino father, castigates his daughter's coworker for "putting ideas into her head" about delaying marriage and attending a university. Teresa, a Portuguese teacher, is shocked when her "American" colleague—a highly Anglicized Latina—accuses her of elitism and rudeness. Teresa is only trying to show proper respect for her colleague's status as a senior teacher when she evades her laughing camaraderie during lunch and after work.

The need for trust and confidentiality, driven by work relationships, makes the segregation of public and private spheres impractical. The new technologically mediated "public space" that intermingles work with home life also confounds the old assumptions about public and private spheres. Old models of containing difference by accentuating status differences still exist in Silicon Valley, but are problematical. The tactics work only when power differences are predictable and stable. New strategies are emerging for the uncertain conditions of cosmopolitan life, but learning to minimize difference and cultivate cultural competencies takes considerable work. All these ways of handling difference are going on simultaneously. How are these myriad strategies integrated to create a functional multicultural community?

Creating a New Culture

Silicon Valley struggles to redefine itself amid the twin processes of identity diversification and technological saturation. There are conflicts.

In the early 1990s, when we first started gathering scenarios of the future of Silicon Valley, people were deeply ambivalent about diversity. On one hand, differences were to be "celebrated," preferably with food and festivals, but they also represented a kind of cobbiness, something just thrown together—a design flaw in the community.

In the projections of technological futures expressed in these early interviews with Silicon Valley educators and engineers, people saw the best future as one in which a single computer/communications platform would prevail: no more messy Apple/PC differences or multiple communications protocols. Phillip, a Smart Valley board member, described the worst future: "[T]he nightmare is that everybody develops independent technologies—there's no coordination, no standards. So we can end up with a lot of interesting and innovative technologies that won't work together, and create a nightmare."

In another version of the pessimistic scenario, one informant added that the people of the worst future "might have multiple different devices that they have to use. They might have to use one device to communicate with their boss, and another device to communicate with the bank, and another device to communicate with their home. All these things will be very annoying to deal with." The "worst" futures were imagined to be a hodgepodge of technologies, all competing and complexifying the market, and myriad separate cultures, each with its own conflicting agenda.

The chaos created by these multiple platforms was seen to be associated with wider social pathologies, such as rampant self-interest, gangs, and market inefficiencies (see Darrah 1996b; English-Lueck 1996; Lewis and Gottlieb 1993: A1, A28). In the "best" futures, though, industry luminaries and teachers alike hoped for a Valley in which cultural division would be subsumed under a single, technologically informed, and enthusiastic citizenry. Interviewees imagine that a good community would have lively ethnic entertainment, exotic food, and upbeat music, but no significant structural differences between ethnic groups. Future Silicon Valley people would speak English—although a few minority voices suggested it would be good to have universal multilingualism. All would share the same accelerated work ethic and respect for capitalism. The best future would feature a single unified technology and a single unified culture—although the specifics reflected the cultural background of the interviewee.

At that time, in the early 1990s, only one person—Angelina, a Center for Employment Training director—voiced the opinion that diversity could be seen not just as a deficit but as an inherent resource in an increasingly global marketplace. This point of view became more common as the decade progressed. As localization and cross-platform engineering created new economic niches and technical opportunities, people began to articulate the idea that the diversity of the Valley uniquely positioned the region's industry to connect to the many other markets and "silicon places" emerging around the world.

The recent "2010 Vision" report on the Valley introduced a similar line of discourse. This vision of the future not only included economic and technical factors, the "spirit of innovation and entrepreneurship," and the natural beauty and climate (all standard defining features of the Valley) but also the "ethnic and racial diversity, and our ability to get along with one another" (Harris et al. 1998: 6B). Cultural capital was seen to combine with natural and technological resources to create a new entity.

Tom talks about how he and his wife were part of the "Technical Gypsy Culture" until he found his niche in Silicon Valley. Working with an international team drawn from widely different professions, he revels in his experience of diversity. He delights in the multicultural environment in which he is raising his children. His statement encapsulates the spirit of ethnogenesis. He says, "I hate to be so immodest, but I feel like we're part of the group that is creating new culture."

This new culture has implications for the rest of the American culture, and the world. Silicon Valley has been an experiment for the twin forces of technological saturation and identity diversity. While the regional culture remains conservative in some realms—people still form face-to-face relationships, and old forms of discrimination are alive and well—SiliconValley has foreshadowed the integration of technology into daily life. It has permitted, even fostered, social innovations around identity, using its own brand of cheerful instrumentality to minimize cultural differences in conditions of deep diversity. What can the world learn from this test case?

CHAPTER 6

Reformatting

Creating Useful Culture

Rush Hour

It is evening in Silicon Valley. People are dispersing in their separate trajectories. Jeff and Aaron, software engineers and family men, are on their way home. Carl is hot on the trail of some information on the internet; he will be joining his family much later this evening. Sheila is at home cooking dinner while her kids are watching videotapes. She hopes that Eric will manage to make it home in time to eat with them, or at least help her with the children's bedtime routine. Norman is getting ready to leave for his evening college class, but he has one more errand to do before he goes. He is helping his grandfather organize a trip back to Taipei. He needs to fax the itinerary he has put together to his Uncle Joseph.

Meanwhile, in Taipei, Joseph Chang drinks his morning Starbucks and daydreams about his upcoming reunion with his father. His mobile phone rings. It is his friend Lao Hsu (old Hsu), calling from a train depot near Hsinchu Industrial Park. They went through the engineering program at Taiwan National University together, and Lao Hsu offers to take Joseph to dinner this evening. Fine, no problem, they will meet at their favorite Sichuan food place in the warren of streets in back of the university. Joseph looks forward to a friendly chat, but Hsu also hopes his friend can find a delicious solution to a technical problem he wants to discuss over their twice cooked pork. This is only fair. Last month Joseph ran into a glitch when he was designing a system integrator for a specific application to be used in wireless communication. As product lead, Joseph was shown the problem by one of his product support engineers. Together they searched the web, chatted with coworkers, and even telephoned counterparts in the company's "R&D" facility in Beijing when e-mail proved problematic. Sometimes communicating with the mainland can be such a hassle. They all brainstormed like mad, but they were stumped. Then Joseph called Lao Hsu and put the question to him. Hsu had run into something similar in a project for a rival company, and had solved the problem. He gave Joseph the bit of information his team needed, allowing Joseph's team to meet an internal deadline two weeks later. After a stretch of long "7–11" days, in which Joseph saw his family awake only at dawn, he took Lao Hsu's family some oranges and brought his sweet little girl a nice doll. Their *guanxi* connection—their history of mutual exchanges and support—had already proven useful, and he was more than happy to maintain it.

Joseph sits in his cubicle facing an open room that is broken into color-coded cubicles. Blue is for product development/engineering, red is for marketing, and purple is for operations. He walks over to his friend Ling and gives her a request for parts. She has a program that surveys the inventory and communicates with vendors for the needed items. Her purple cube is decorated with pictures of her two nephews and her parents. She also has a stuffed animal gracing the top of her monitor.

Joseph walks out to the "smoking area" on the balcony. It is near the toilets and everyone jokes about how it functions as the informal conferencing area of the company. He is thinking about quitting smoking, but as he lights a cigarette he muses that then he would have no excuse to chat with people in other product lines and divisions and find out what is really happening in the company. A fellow from another product team tells him about a wonderful web site he visited where he can find a hotel, book a flight, and even get all his maps for an upcoming trip to Japan. Another man tells him about a funny incident involving an exchange of gifts when he went to Beijing. Some of the people in the organization there had heard that he was highly placed in the company, and so had bought him some nice *guanxi* gifts, hoping to establish a connection. It was so funny watching them trying to hide their discomfort when they realized that his position was roughly horizontal to theirs, and that he was not an influential person at all. The Beijing group really wished they hadn't wasted their resources on someone who couldn't help them, but they had to be graceful about it. Mainlanders! Joseph stretches and goes back to work. The rumble of the traffic and the train grows weaker as he moves inside the building.

On the train is Cynthia Lin. She rapidly steps off a brown line train on the MRT Rapid Transit, having left her home in Mucha, a suburb of Taipei. Cynthia works in the heart of Taipei for a financial firm whose tendrils reach throughout Asia and to the United States. Cynthia needs to understand the American way of business and teach it to her Taiwanese clients. She often has to be a cultural interpreter. She has to understand the American concept of "business model": that is, how people intend to make money. When some high-tech people came to her with a proposal for asking for American venture capital, she had to tell them they needed to make a clear case explaining how they would make money. Web pages don't just make money automatically, she told them. Are they going to make a product that they will distribute? Are they going to sell their services to a big international software company? Then, this other company would be their client. Sometimes these young "gifted kids" just don't understand these things.

Cynthia speaks fluent English, and so often interacts with Americans and Europeans. She feels a bit resentful. In a fairer world, more people would speak Chinese. Instead of working so hard to learn English she could have been learning that much more about finance. Her grandfather had been educated by the Japanese, when Taiwan was a Japanese colony, and she grew up learning all sorts of things from him about Japanese language and culture. Sometimes she passes that information to her clients as well. She has learned how to negotiate with Hong Kong Chinese, who, in spite of their prosperity, are really very different from the stodgy older generation in Taiwan who don't always understand the pace and requirements of global business.

Cynthia is twenty-nine, unmarried with no children, living with her parents. She has three siblings—two younger sisters and an older brother. She and all of her siblings are armed with mobile phones. There is a widespread feeling that mobile phones are cheaper than wired, but as an astute business woman, Cynthia knows it isn't really true. However, they are convenient. She used e-mail to stay in touch with her brother when he lived in the United States.

He worked for a Silicon Valley tech company for a time. He could have stayed, but he missed his family and he didn't really like having to speak English all the time at work with a bunch of foreigners who acted as if his fluent, but accented, English was a reflection on his intelligence. But he is home now, a senior manager in a specialized Chinese language applications software firm in Hsinchu. There he can use the connections he made in Silicon Valley, but he is more comfortably surrounded by familiar sights and smells, and his friends and family. They talk together on the phone at least four times a week, and the whole family gets together for dinner almost every weekend. When Cynthia's brother lived in California, she e-mailed him, as she does her friends from college who now live in New York. She has convinced her father to get a computer and an internet service for their home. Her father hardly touches the computer, but her mother and sisters love it.

Her mother has dabbled in art history since her retirement, and she loves the internet because "she can pretty much go everywhere." Cynthia notes that her mother "can surf to the web site on the Louvre and see what's going on there. Even though she doesn't understand French, at least she can look at the pictures." She likes feeling "connected." Cynthia's youngest sister is doing graduate studies at the university now and uses e-mail to connect with classmates who graduated and went overseas. Sometimes Cynthia needs her sister to do some shopping for her and gives her a call from work. However, she would never dream of e-mailing her sisters with a request. Imagine talking to your own sister over e-mail here in Taiwan! Isn't that a bit cold?

As the train speeds along, Cynthia gazes out the window, momentarily thinking about the direction of her own life as her image is dimly reflected on the glass. Her friend at work is always talking about going to the United States. She calls Silicon Valley "the Center of the World." Sigh. Cynthia might want to go to America for a year or two—maybe. She has too many obligations to leave for very long. Her parents need her, as do her sisters. She hopes to get married soon. She can't just walk away from all that. In any case, she has traveled enough to know that she likes living in Taiwan. It has just the right blend of the old traditional ways and the new global excitement. Working in Silicon Valley might rob her of her warm relationships and cause her to lose her Chinese self.

Silicon Valley Cultures Revisited

Throughout this book I have been looking at the double helix that outlines Silicon Valley—technological saturation and identity diversity. The region's "critical mass" of technology permeates people's lives completely, reframing the choices that people make every day. Technology facilitates the creation of networks—a form of social organization that is reshaping social institutions from family to civic politics. Device-mediated networks of people—local and transnational—compose a new electronic public space in which cultural processes are enacted and created. The saturation of technology changes the very metaphors and problem-solving approaches people use. Technology changes the cultural responses of daily life and reshapes people's identities. In turn, cultural identity and cultural competence become tools, ways people can engineer their communities. The area is a laboratory for "deep diversity," reflecting the contribution of dozens of groups. Astronomical numbers of cross-cultural interactions give rise to a wide variety of strategies for dealing with the resulting complexity.

In this final chapter of the book, I will systematically review the "culture" of Silicon Valley—its distinctive political economy, innovative forms of social organization, and characteristic beliefs and symbols. The Valley is a technological haven, home to a specialized economy that depends on global connections. It is the prototype for a technologically saturated community, in which devices have become deeply integrated into the lives of the people who produce them. Silicon Valley pioneers forms of social organization that have wide-ranging implications for other communities of technological producers.

Work has become the most important domain in life. Personal networks transcend organizational boundaries and confound the demarcation of clear and simple cultural groups. Silicon Valley is more than an incubator for newborn technologies; it incubates new cultural philosophies as well. Placing technological and organizational metaphors at the center of community life reframes the meaning of social life.

The primacy of work, the elevation of efficiency to a commandment, and the benign instrumentality with which people network all point to an ideology of pragmatism. That pragmatism, a distillation of a particularly American philosophy, governs civic life and interpersonal interactions. Pragmatism underlies the ethos that manages—and sometimes minimizes—cultural differences and allows diverse people to function productively together.

Finally, in the last section of this chapter, I will discuss the lessons that can be learned in Silicon Valley's cultural laboratory. Is Silicon Valley a place it would be desirable to replicate, or is it a bad example, one to be avoided? Is it a bellwether, showing us an early glimpse of cultures to come? Whether one lives in a "silicon place"—such as Austin, Bangalore, Dublin, or Taipei—or simply in a community in which consumer electronics and incoming migrants are challenging the traditional ways of life, Silicon Valley offers an image that people around the world can compare with themselves.

Technocracy Revised

The saturation of a community's culture with technology is a process that takes place at several levels. At one level, the regional political economy is dominated by high-technology industries. This economic base is volatile, characterized by spectacular booms and crashes. The rapid pace of technological innovation engenders a more general velocity that filters into corporate policies and individual lives. Work is multifaceted, involving many interdependent components, human and material. High-tech work is also global, drawing on nodes of human talent, capital, and resources around the planet.

Technologically saturated communities attract and hold more educated people and reinforce the value of wealth as demand increases for such commodities as real estate and services. High rents and shortages of skilled labor are characteristic of "silicon places." New political structures develop to support the new lucrative economy. Joint Venture in Silicon Valley, and Forfas in Ireland, are quasi-political structures designed to facilitate high-tech economic growth (see McBrierty and Kinsella 1998: 148–49; Forfas 2000). Forfas, an organization that promotes Ireland's commercial application of science and technology, affirms the familiar values of "innovation," and "competitiveness." However, Forfas also emphasizes Ireland's connection to a broader European social framework, and recognizes the wide range of low-tech industries that are also essential to Ireland's economy (Forfas n.d.). Elements of Silicon Valley's technologically saturated lifeway may be adopted, rejected, or reinvented to suit the needs of the local culture.

Discussions of Silicon Valley inevitably lead to one topic—can it be reproduced? But that question disguises another inquiry: Can other technologically identified communities flourish as Silicon Valley has flourished? Can Taipei become another Silicon Valley? Can Dublin? Can Austin? These questions imply a search for the secret formula for Silicon Valley's economic success. Take four universities, multiply by two industrial parks, and add a pinch of good weather. Unfortunately, these attempts to replicate the Valley's experience do not always account for the density and diversity of people, networks, connections, and industries that it includes. Historians have pondered the legacies of industrial innovation as far back as the early part of the twentieth century (Ignoffo

1991), and urban analysts have earlier noted the distinctive advantage of flexible networks (Saxenian 1994).

Instead of thinking about cloning Silicon Valley's success, a more realistic approach might look at how other "silicon" locales of the world can *use* the industrial infrastructure and social innovations developed in Silicon Valley to create their own niches. San Francisco's Multimedia Gulch has used Silicon Valley's technology and capital to do so. Taiwan's Taipei/Hsinchu corridor is developing a niche by localizing existing Silicon Valley products for Asian use, while Silicon Valley companies become advanced training grounds for a new generation of Asian indigenous entrepreneurs. "Silicon" communities are not entirely discrete from each other, but rather, intertwined in a global network that colonizes existing infrastructure and creates new opportunities.

Not every community wants to be as technologically saturated as Silicon Valley. Even though its citizens are sophisticated in the use of electronic devices, Hong Kong is a center of finance, not technology. Hong Kong's companies facilitate relationships throughout Asia, but they do not produce the technology they market. New York, Tokyo, and London are complex financial centers, with pockets of technological saturation, but the production of technology is only one small part of their economies, not the dominant one. The presence of technology in such global centers is a reflection of consumer use.

Technology penetrates the lives of people in many places—Des Moines, Paris, and Shanghai. In their use of everyday technologies, Silicon Valley people may not be so different from people in other places. But producers of technology use the devices differently from less involved consumers. Widespread technical literacy and wholesale technical enthusiasm have an effect on the way even mundane devices are used. People choose tools—devices and services—to shape their relationships. Artifacts are given a special meaning by the fact that the people who are buying them are also responsible for producing them. Because people have to work interdependently to create, construct, and distribute their products, work is dominated by communication.

In the complexity of the modern creative workplace, however, people work at different rhythms and paces. When work is global, coworkers do not all occupy the same time zone, and the need for workers to juggle the rhythms of distant colleagues shapes the choice of tools, requiring the extensive use of asynchronous communication. Interdependence, however, also requires a depth of familiarity and trust that cannot easily be established asynchronously. As a result, people use a variety of media to maintain contact, to sustain and create relationships that endure through divorce, downsizing, and school redistricting. These tools enable people to extend their power and to resist the power of others.

This nuanced use of tools is possible only in a social environment saturated with technological devices. People in Silicon Valley talk about the devices they use; they revel in "technolust." That is, in part, because they are a community of technological producers. To deny the delight of technology would undermine the very raison d'être for the community. It does not mean that the rest of the world will resemble Silicon Valley, any more than the nearly universal use of automobiles meant that every place in the United States has become another Detroit.

Silicon Valley culture, characterized by the penetration of work into personal, home, and community life, may foreshadow the future of other technologically oriented communities and influence their choices. When Gale, a Dublin woman, talks about her work in a high-tech organization, she sounds as if she could be in the heart of Silicon Valley:

The downside of a mobile [cell phone], voice mail, PC, or the laptops, is that people think that you are online and available twenty-four hours a day. I've noticed that recently. I was getting a call last week at a quarter to six in the morning from Australia. And it was great, you know, "Good morning! This is Sally. How are you?" And I go, "I'm still in bed and it's dark outside. . . ." You're up very early for Australia, and you're working very late at night for California, and so on. . . . Sometimes you get calls right up to seven, eight, nine, ten [at night]. If you've been going [since] early in the morning, you're not the freshest. No human being could be. It's very hard to be at your desk twenty-four hours a day within a global organization. . . . I've noticed in the past couple of weeks you're just working everywhere . . . you might be in a restaurant, or you might be in the supermarket, and you're expected to stand there and chat about these various issues while you're picking up your milk and your cheese and all that. . . . It's great to have the flexibility, but the downside is you are expected to be on call twenty-four hours a day then. . . . We never have the situation where people refuse to go somewhere, or refuse to do anything. They're quite open to working very hard for the company. We don't have overtime here. People just do it out of love for what we're doing. And because we believe in the vision of where the company's going, and that's great. The people—I mean we've got developers who are in here until three or four o'clock in the morning.

Reilly, however, makes it clear that technological saturation in Ireland goes only so far. His brother and father are "fencers-farmers-undertakers"—they have no e-mail at all and exchange only an occasional phone call. His sister works in an office and does have e-mail, but it is an open system that can be read by everyone in the firm. Reilly will occasionally e-mail her in Irish, which nobody else in her office speaks, just for the sake of privacy. He also will use e-mail to write a friend in New Zealand, another techie. But those are the exceptions. Even though he works in a high-tech company, not everyone in his world embraces the tools. His experience is similar to that of people in Silicon Valley, but it retains distinctive local characteristics. Even a "silicon place" does not require the complete reproduction of Silicon Valley culture.

Organizing for Complexity

People in Silicon Valley create social organizational tools to manage complexity. People reframe work to make it the dominant lens through which problems are seen and resolved. Work is often the reason for coming to, or staying in, Silicon Valley. Work practices and beliefs are expressed in styles of parenting and recreation. Work organizations promote a technological problem-solving strategy that refashions local politics. Work becomes a model for reorganizing social relationships, and community life is centered around the workplace and work-related interactions. The domains of work, family, church, community service, and personal interests intertwine. Social life is increasingly lived in networks whose membership cuts across traditional groupings. In an Asian-American church in which 80 percent of its members are engineers, is the group organized around religiosity, technical prowess, or ethnic identity? The networks formed in Silicon Valley interweave all of the above.

Silicon Valley people use a cheerful opportunism and instrumentality to create and maintain connections. The creation of networks is enmeshed with the technologies that link people into clusters of communication. Some links are deep and need to be used only rarely to be meaningful. "Friends of the heart" may not appear often, but the trust placed in them is deep. Other connections may be shallow but are renewed in daily contact. A friendly e-mail establishes contact with a project teammate, but such friends are "friends of the road." Relationships are created, maintained, and molded with technological devices. The public space in which culture is constructed is technologically mediated, as focused as a telephone call or as diffuse as a web site.

There are contradictions built into this device-mediated social networking system. Beyond the obvious social problems generated by rapid economic growth—escalating cost-of-living, environmental degradation, and a perilous rich-poor gap—there are other, more subtle dangers. At the individual level, people rely on social networks that are based on trust. Whether the need is to solve a technical problem, find a suitable nanny, or get the next job, networks provide the real social web that prevents free-fall, and those networks must be sustained by personal contact. Ironically, however, while asynchronous communication—messaging and e-mail—apparently grants unlimited access, it may interfere with building trust. The devices work well when trust is already in place but may create problems when the trust is still being established.

As if device-mediated interaction were not complex enough, people in Silicon Valley cannot assume that the others in their world share the same ideas of work, power, and community. Personal contact can also be problematic when multiple cultures are involved. Where corporate culture features a "just-in-time management" style, it is easy to assume that one knows what is expected, only to find that "on time" may mean something different to someone from another culture or class.

People in Silicon Valley live amid a "deep diversity" that requires that they invent new and flexible social identities. There are different levels to this diversity. Each person may assume identities born from a variety of sources—national and regional origin, ethnic affiliation, professional training and personal interest. The number of possible interactions within and among identity groupings is astronomical. People manage that complexity in an enormous variety of intricate ways. Sometimes cultural differences are maximized—where stable power differences persist—and evoke such traditional responses as racial exclusion and class division. People carve out comfort zones of similarity. Other less predictable situations clearly require strategies that minimize cultural difference. People construct buffers to the inevitable violations of their unconscious expectations that occur when different cultures interact. The primacy of work and technology form an alternative framework that can be used to override other differences.

In other "silicon places" the experience of cultural diversity may be different, and cultural identities may be differently constructed. Dublin's burgeoning high-tech industry is not quite monocultural; it embraces other Europeans, as well as Americans and Australians, many of whom are "returning swans"—former expatriates who are Irish or of Irish descent. But Dublin lacks some of the cultural tools needed to deal with the Indians and Chinese who are beginning to colonize its high-tech economy. In spite of the fact that Hong Kong includes hundreds of thousands of foreign-born expatriates (some working as servants, and others members of the social elite), it does not view itself as multicultural, but as Chinese—and, specifically, Hong Kong Chinese (English-Lueck 1995). Taiwan, on the other hand, consciously uses its cultural and linguistic connections to the People's Republic, and its historical ties to America and Japan, as cultural bridges.

These differences are not just the product of individual choices. Silicon Valley people tend to forget that social institutions are not just commodities but also agents of change. The infrastructure of the Valley has been transformed to facilitate the saturation of technology and to create a place for the cultural tools to flourish. Where else do secondary schools offer advanced training in biotechnology? Community is seen as a platform for multiple business, ethnic, and cultural communities, and different political constituencies—be they libertarian technocrats or gay advocates. Imbuing civic discourse with the language of technology and work offsets potentially balkanizing cultural differences. How successful this new infrastructure might be is not yet known, but currently it provides a social organization that can accommodate both technological saturation and identity diversity.

Many community-wide solutions to problems depend on making work and technology the first principle. Innovation, efficiency, and constant reinvention are hallmarks of Silicon Valley's civic life. But this may create contradictions at the community level. As an adaptation to complexity, making work the primary organizing principle has been successful, but is it sustainable? High-technology industries are the most volatile sectors of the economy. When the sector expands, it "booms," but it can fail catastrophically. It is a vulnerable basis for community.

The Etiquette of Pragmatism

When people move to Silicon Valley from Stockholm or Portland, they do not come solely for the technology but also for the "culture" that produces it. The ethos that generates innovation thus becomes the exclusive commodity of Silicon Valley. What is that ethos? Created in response to the requirements of identity diversity, and discussed in the rhetoric of technology, Silicon Valley embodies the virtues and vices of pragmatism. When Clifford Geertz wrote about life in Bali, he painted a scene in which individuals acted in "a haze of etiquette, a thick cloud of euphemism and ceremony" (1990: 121). In Silicon Valley, the atmosphere is imbued with the language of pragmatism; the etiquette is based on cultural instrumentality.

The pervasiveness of technological metaphors in daily and civic life elevates "efficiency," "innovation," and "invention" to the paramount virtues. These concepts are then applied in new contexts. Negotiating the highway system is described in terms of "driving efficiency," and "spiritual reinvention" is promoted by a favorite natural retreat. Efficiency and flexibility, staples of Silicon Valley rhetoric, are deeply pragmatic values, and the pragmatic language of engineering reshapes the imagery of daily life.

Pragmatism, as a philosophy, was a nineteenth- and twentieth-century American movement that viewed truth as a function of experience and practical outcome. Principles are not absolute, but emerge from working hypotheses. The mind is an instrument to be used for the fulfillment of practical goals. Through the historical replication of American notions of public education and democratic institutions, the idea now has a global audience. American pragmatism has combined with similar indigenous notions—for instance, elements of Chinese Confucian thought—to promote instrumental reasoning. Classical Chinese *guanxi* obligations, which bind people together in an ongoing state of reciprocal favors, are ultimately *useful* relationships. *Guanxi* ties are invoked to find the freshest vegetables or to get your children into the best universities. In the world of technological producers, this emphasis on usefulness underpins the creation of a technocratic "meritocracy" in which those with technical ability become the social elite.

This philosophy of pragmatism is visible in the etiquette that governs intercultural interaction in Silicon Valley. While racism and cultural separatism certainly exist, as artifacts of contemporary American life, such attitudes interfere with creating working networks. Civility and tolerance prove more effective in building relationships. Because the technical "meritocracy" can produce sudden shifts in power hierarchies—today's underling might be tomorrow's boss—people are careful to be pleasant to each other. While recognizing that one person's idea of what is pleasant behavior may differ from another's, people will try to be flexible and forgiving. A pact of tolerance for mutual cultural clumsiness is a powerful cultural innovation.

Where work and technology cannot provide the basis for an alternative discourse, the tools for minimizing difference are also reduced. The success of Silicon Valley depends on being able to exploit the diverse people and ideas that pass through. Some of the people we interviewed thought that the tolerance and flexibility inherent in multicultural adaptation train the mind itself to be more "open." That openness is linked to creative, innovative thinking, the chief asset of the region. Being prepared for the ambiguity of human interaction requires a creativity that can be transferred to one's work. When asked to consider the impact of diversity, Harry, a computer scientist, opines colorfully, "It might actually help just having weird, wacky backgrounds so that you're more prone to challenging established conventions."

However, the pragmatic "aura" that may grant Silicon Valley a measure of success comes at a price. An ethos of cultural flexibility may encourage creativity, but informants also expressed concern about the difficulty in making meaningful connections, or in sustaining a "deep" sense of belonging. One of the consequences of tolerating alternative cultures is that it becomes more difficult to maintain an unambiguous belief in the inherent naturalness of one's own birth culture. The loss of security in tradition is keenly felt. As work permeates the home, and private and public spaces interact and merge, cultural flexibility is demanded in the most intimate relationships. Home life can no longer be treated as a sacred preserve of one's cultural traditions. This level of cultural flexibility may be more than some people are willing to practice.

The cultural tradeoffs for the Silicon Valley ethos are hard to calculate. Many of the advantages and costs are subtle; they are hard to detect without considerable reflection. Luke, an instructional designer and intercultural trainer, appraises the situation:

The pace here happens so fast you can really only observe it as it's happening, without really being able to tell what the implications are necessarily, because it is too fast. It's like standing in front of the waterfall trying to count the raindrops. You can kind of see them twinkling as they go by. . . . And I don't know if that's good or bad. I don't know.

Luke goes on to comment that while the old local cultural nuances and individual traditions are being lost in the haze, technology also allows cultural knowledge to be preserved and communicated as never before in human history. How can we anticipate the future of the culture that is being created in Silicon Valley?

Battling the Metaphors

Is Silicon Valley a bellwether, a glimpse into the future lives of our children? The community has marketed itself as the "land of tomorrow," but that rhetoric may not be borne out. Silicon Valley's intense technological saturation makes it an inappropriate model for many other communities—even for other "silicon places." The Valley is home to thousands of high-tech production firms. In every cafe and pub, one hears deeply technical discussions. The political economy of Silicon Valley cannot be widely duplicated.

The people we interviewed repeatedly called Silicon Valley "a technological Mecca," characteristically blending the languages of religion and the workplace. Internet gurus from Taiwan and software security specialists from Ireland talked as if a sojourn to Silicon Valley was a necessary rite of passage into a high-tech world. Certainly, religious metaphors are appropriate, in that Silicon Valley workers view themselves and their work with near-missionary zeal. Enabling efficiency, communication, and connection are considered venerable, almost spiritual, goals. But calling Silicon Valley a Mecca has other, more foreboding, implications.

After the introduction of Islam to India, Mecca became an unintentional repository for epidemic diseases. Pilgrims brought the plague from India on their hadj to Mecca, from where the disease was transmitted by other pilgrims to other parts of the Moslem world, including Southern Europe. What cognitive "pathogens" might be transmitted back to Taipei and Dublin? Community reinvention? Pervasive networking? Organizational hopscotch? Cultural instrumentality? The primacy of efficiency? People may come to Silicon Valley as pilgrims to partake of its technological and entrepreneurial banquet, but come away with other unintentional cultural accouterments.

Finally, Silicon Valley's name is itself a metaphor. A polished silvery wafer of silicon, ready for the imprint of sophisticated circuitry, is as reflective as a mirror. As the people of various communities hold Silicon

Valley up to themselves, they see the wafer, but also see themselves. People from the outside can use the image to ask, "How does our community, our place, compare to Silicon Valley?" Do we envy its technologically inspired prosperity, or do we shudder at the pace of life? Do we admire its bright multicultural mosaic, or retreat in horror from its matter-of-fact instrumentality? Silicon Valley allows us to reflect on the choices that are made when a community is technologically saturated. Some people will romanticize high-technology work and be attracted to its creativity. The cultural stimulation and demands for intellectual flexibility will attract some. Others will be drawn by the promise of wealth. But not everyone wants to live in Silicon Valley, or even another "silicon place" that partially replicates it. The choices the people of the Valley have made, individually and collectively, have consequences. Viewing life through technological metaphors risks narrowing one's worldview to exclude other alternative possibilities. Reinvention comes at the cost of just "being." The sheer time requirements of maintaining work and social networks—some weak, others intense—that span social differences and extend around the globe, can be overwhelming. To maneuver through these time constraints, people have redefined their homes and social lives as places of production, true cottage industries. Not everyone views that solution with comfort.

Silicon Valley shows us what life can be like when the village plaza is decorated with internet graphics and the fabric of social life is woven from digital threads that link people together. Technological saturation restructures daily life so that boundaries between work, home, and community activity are softened and even dissolved. It is one illustration of what can happen when a new vocabulary and a realignment of community interests create civic life under late capitalism. Silicon Valley shows us some of the ways we can adapt to ethnic and cultural diversity that promise to become more intense everywhere in the years to come. Some mechanisms for managing intercultural interaction are drearily familiar—racism, classism, and ghettoization. Other strategies point to a new way to actively engage "deep diversity"—to learn to be uncertain about the impacts of one's behavior, to be wary of giving offense, and to learn to be inured to cultural behaviors that differ from your expectations. All of these defining features of Silicon Valley have consequences; Silicon Valley provides a mirror to reflect on them.

References Cited

Akizuki, Dennis
    1999a Buddhism in Blossom. San Jose Mercury News, May 23: 1A, 10A.
    1999b Taiwan Ties. San Jose Mercury News, August 1: 1A, 22A.
Appadurai, Arjun
    1996 Modernity at Large: Cultural Dimensions of Globalization. Minneapolis: University of Minnesota Press.
    2000 Grassroots Globalization and the Research Imagination. Public Culture 12(1): 1–19.
Baba, Marietta
    1999 Dangerous Liaisons: Trust, Distrust, and Information Technology in American Work Organizations. Human Organization 58(3): 331–46.
Bailey, Brandon
    1998 Hunger Figures Rise in Bay Area. San Jose Mercury News, March 10: 1A.
Baldassare, Mark
    2000a Californians and Their Government. September 2000. San Francisco: Public Policy Institute of California.
    2000b Californians and Their Government. October 2000. San Francisco: Public Policy Institute of California.
Barley, Stephen
    1988 On Technology, Time and Social Order: Technically Induced Change in the Temporal Organization of Radiological Work. In Making Time: Ethnographies of High-Technology Organizations. Frank Dubinskas, ed. Pp. 123-69. Philadelphia: Temple University Press.
Barsook, Paulina
    2000 Cyberselfish: A Critical Romp through the Terribly Libertarian Culture of High-Tech. New York: PublicAffairs.
Barth, Fredrik
    1969 Ethnic Groups and Boundaries. Boston: Little, Brown and Company.
Bassman, Theda
    1997 Treasures of the Navajo. Flagstaff, AZ: Northland.
Benedict, Ruth
    1989 Patterns of Culture. Boston: Houghton Mifflin.

Bennett, Noël, and Tiana Bighorse
   1997   Navajo Weaving Way. Loveland, CO: Interweave Press.
Bernal, Martha, George P. Knight, Kathryn Ocampo, Camille Garza, and
      Marya Cota.
   1993   Development of Mexican American Identity in Ethnic Identity: For-
         mation and Transmission among Hispanics and Other Minorities.
         Martha Bernal and George Knight eds. Pp. 31–46. Albany: State
         University Press of New York.
Bohannan, Paul
   1995   How Culture Works. New York: Free Press.
Brislin, Richard
   2000   Understanding Culture's Influence on Behavior. 2d ed. San Diego:
         Harcourt College Publishers.
Bronson, Po
   1999   The Nudist on the Late Shift. New York: Random House.
Bureau of Inter-American Affairs
   1998   Background Notes: Suriname. Electronic document. http://www.
         background_notes/suriname_0398_bgn.html.
Cain, Bruce, Jack Citrin, and Cara Wong
   2000   Ethnic Context, Race Relations, and California Politics. Public Pol-
         icy Institute of California. Electronic document. http://www.ppic.
         org/publications/ppic137/ppic137.pdf.
California Department of Education
   2000   Enrollment in California Public Schools by County and by Ethnic
         Group, 1999–2000. Electronic document. http://data1.cde.ca.gov/
         demographics/reports/statewide/.
   2001   San Jose Primary Metropolitan Statistical Area (Santa Clara County,
         California) Labor Force and Industry Employment. Electronic docu-
         ment. http://www.calmis.ca.gov/file/lfmonth/sanj$pr.txt
California Employment Development Department
   2001   Civilian Labor Force, Employment and Unemployment. Santa Clara
         County. Electronic document http://www.calmis.cahwnet.gov/file/
         lfhist/sanj$hlf.txt.
Carrico, J.
   1996   Cultural Diversity Training: Corporate Stratification of Cultural Di-
         versity? M.A. thesis. San Jose State University.
Casey, Catherine
   1995   Work, Self and Society: After Industrialism. New York: Routledge.
Cassidy, Mike
   1997   High Tea, High Tech; Local, Galactic Causality; Whatever—It's
         Fry's. San Jose Mercury News, July 16: 1E.
Castells, Manuel
   1996   The Rise of the Network Society. Cambridge, MA: Blackwell Pub-
         lishers.

Castells, Manuel, and Peter Hall, eds.
  1994   Technopoles of the World: The Making of Twenty-first-century In-
          dustrial Complexes. New York: Routledge.
Castilla, Emilio, Hokyu Hwang, Ellen Granovetter, and Mark Granovetter
  2000   Social Networks in Silicon Valley. *In* The Silicon Valley Edge.
          Chong-Moon Lee, William Miller, Marguerite Gong Hancock, and
          Henry Rowen, eds. Pp. 218–47. Stanford: Stanford University Press.
Cha, Ariana
  1997   Immigrant Brain-drain. San Jose Mercury News, July 28: 1A, 16A.
Chen, Anne
  1997a Passage from India. San Jose Mercury News, August 10: 1A, 28A.
  1997b Schwab Invests in Office Geared to Chinese-Americans in Silicon
          Valley. San Jose Mercury News, July 17: 1C, 3C.
Christie, Linda
  1997   California Native American College Students' Experience: An Eth-
          nographic Study. M.A. thesis. San Jose State University.
Cohen, A.
  1994   Self Consciousness: An Alternative Anthropology of Identity. New
          York: Routledge.
Collaborative Economics
  2001   Unfinished Business: Women in Silicon Valley Economy. Project Re-
          port. Palo Alto: Collaborative Economics/Community Foundation
          Silicon Valley.
Commission on the Advancement of Women and Minorities in Science, Engi-
          neering and Technology Development
  2000   Land of Plenty: Diversity as America's Competitive Edge in Science,
          Engineering and Technology. September 2000. Report of the Con-
          gressional Commission on the Advancement of Women and Minori-
          ties in Science, Engineering and Technology Development. Electronic
          document. http://www.nsf.gov/od/cawmset/report/cawmset_report.
          pdf.
D'Andrade, Roy
  1984   Cultural Meaning Systems. *In* Culture Theory: Essays on Mind, Self,
          and Emotion. Richard Shweder and Robert Levine, eds. Pp. 88–119.
          New York: Cambridge University Press.
Darrah, Charles
  1994. Skill Requirements at Work. Work and Occupations 21: 64–84.
  1995   Workplace Training, Workplace Learning. Human Organization 51:
          264–73.
  1996a Learning and Work: An Exploration of Industrial Ethnography.
          New York: Garland Press.
  1996b Community and Collaboration in a Value-Added Valley. Paper pre-
          sented at the Annual Meeting of the American Anthropological As-
          sociation, San Francisco, Nov. 20.

1997   Listening to the Voices of Diversity. Unpublished report. Department of Anthropology. San Jose State University.

Darrah, Charles, J. A. English-Lueck, and James Freeman
2000   Living with Technology. *In* Anthropology and Middle Class Working Families: A Research Agenda. Mary Margaret Overbey and Kathryn Dudley, eds. Pp. 40–43. Arlington, VA: American Anthropological Association.

Darrah, Charles, J. A. English-Lueck, and Andrea Saveri
1997   The Infomated Households Project. Practicing Anthropology. Fall 1997.

Dedera, Don
1975   Navajo Rugs. Rev. ed. Flagstaff, AZ: Northland.

Delbecq, André
1994   Innovation as a Silicon Valley Obsession. Journal of Managerial Inquiry 3(3): 266–75.

DeVol, Ross
1999   America's High-Tech Economy: Growth, Development, and Risks for Metropolitan Areas. Santa Monica: Milken Institute.

Dew, Edward
1990   Suriname: Transcending Ethnic Politics the Hard Way. *In* Resistance and Rebellion in Suriname: Old and New. Studies in Third World Societies. Publication 43. Gary Brana-Shute, ed. Pp. 189–212. Williamsburg, VA: College of William and Mary.

Dorgan, Michael
1997   Costa Rica Gets on High-Tech Map. San Jose Mercury News, August 10: 1D, 6D.

Eng, Sherri
1997   Letter Perfect. San Jose Mercury News, November 5: 6G, 8G, 9G.

English-Lueck, J. A.
1995   The Difference Engine: Creating Identity in Silicon Valley. Paper presented at the Annual Meeting of the American Anthropological Association, Washington, D.C., Nov. 15.
1996   Retrospective on Tomorrowland: Visions of the Future in Silicon Valley. Paper presented at the Annual Meeting of the American Anthropological Association, San Francisco, Nov. 20.
1997   Chinese Intellectuals on the World Frontier: Blazing the Black Path. Westport, CT: Bergin and Garvey.
2000a   Work, Identity and Community in Silicon Valley. Final Report for Award 9810593. National Science Foundation.
2000b   Silicon Valley Reinvents the Company Town. Futures 32: 759–66.

English-Lueck, J. A., and Charles Darrah
1999   Silicon Work: Tales of Trust from Silicon Valley, Bangalore, Taipei and Dublin. Paper presented at the Society for Philosophy and Technology 11th Biennial International Conference, San Jose, July 16.

English-Lueck, J. A., Charles Darrah, and James Freeman
  2000   The Silicon Valley Cultures Project. Karl Lueck Designs. Electronic document. http//www.sjsu.edu/depts/anthropology/svcp/.
Erasmus, Melanie
  1996   Immigrant Entrepreneurs in the High-Tech Industry. *In* Reframing the Immigration Debate. Bill Ong Hing and Ronald Lee, eds. Pp. 179–94. Los Angeles: Leadership Education for Asian Pacific (LEAP) and UCLA Asian American Studies Center.
Evans, Grant, and Maria Tam, eds.
  1997   Hong Kong: The Anthropology of a Chinese Metropolis. Honolulu: University of Hawai'i Press.
Ewell, Miranda
  1997   Sweet Sorrow Tech-Style. San Jose Mercury News, June 17: 1A, 8A.
Feenberg, Andrew
  1991   Critical Theory of Technology. New York: Oxford University Press.
Forfas
  n.d.   Science Technology and Innovation in Ireland. Dublin: Department of Enterprise and Employment, Office of Science and Technology.
  2000   Forfás. Electronic document. http/www:forfas.ie/mainpage.htm.
Freeman, J. M.
  1989   Hearts of Sorrow: Vietnamese American Lives. Stanford: Stanford University Press.
  1996   Changing Identities: Vietnamese Americans 1975–1995. Boston: Allyn and Bacon.
Fricke, Tom
  2001   Taking Culture Seriously: Making the Social Survey Ethnographic. Center for the Ethnography of Everyday Life Working Paper 022–01. Ann Arbor: University of Michigan, An Alfred P. Sloan Center for the Study of Working Families.
Geertz, Clifford
  1990   The Balinese Cockfight as Play. *In* Culture and Society: Contemporary Debates. Jeffrey Alexander and Steven Seidman, eds. Pp. 113–21. New York: Cambridge University Press.
  2000   Available Light. Princeton: Princeton University Press.
Gillmor, Dan
  1997   Why High Tech Embraced Ireland. San Jose Mercury News, July 27: 1D, 2D.
  1998   Growing Links between Israel, Silicon Valley. San Jose Mercury News, May 10: 21A.
Glatstein, Leslie
  1994   Smart Valley, Inc.: A Cooperative Effort to Develop an Information Highway for Silicon Valley. M.A. thesis. San Jose State University.
Ha, Oanh
  1999   Bridging the Commercial Divide. San Jose Mercury News, July 26: 1E, 12E.

2001   Brokerages See Asian-American Dividends. San Jose Mercury News, January 11: 1C, 4C.

Hakken, David
1993   Computing and Social Change: New Technology and Workplace Transformation, 1980–1990. Annual Review of Anthropology 22: 107–32.

Harmsen, W. D.
1985   Patterns and Sources of Navajo Weaving. Denver: Harmsen Pub.

Harris, Jay, et al.
1998   Envisioning the Valley. San Jose Mercury News, October 11: 1P, 6P.

Hayes, Dennis
1989   Behind the Silicon Curtain: The Seductions of Work in a Lonely Era. Boston, MA: South End Press.

Hendricks, Tyche
1997   Shedding Light on Islam. San Jose Mercury News, December 27: 1E, 3E.

Henton, Doug
2000   A Profile of the Valley's Evolving Structure. In The Silicon Valley Edge. Chong-Moon Lee, William Miller, Marguerite Gong Hancock, and Henry Rowen, eds. Pp. 46–58. Stanford: Stanford University Press.

Hochschild, Arlie
1997   The Time Bind. New York: Metropolitan Books.

Hoogbergen, Wim
1990   The History of the Surinamese Maroons. In Resistance and Rebellion in Suriname: Old and New. Studies in Third World Societies. Publication 43. Gary Brana-Shute, ed. Pp. 65–102. Williamsburg, VA: College of William and Mary.

Hossfeld, Karen
1988   Divisions of Labor, Divisions of Lives: Immigrant Women Workers in Silicon Valley. Unpublished dissertation. Department of Sociology, University of California, Santa Cruz.

Hurtado, Aida, Jaclyn Rodriguez, Patricia Gurin, and Janette Beals
1993   The Impact of Mexican Descendants' Social Identity on the Ethnic Socialization of Children. In Ethnic Identity: Formation and Transmission among Hispanics and Other Minorities. Martha Bernal and George Knight, eds. Pp. 131–62. Albany: State University Press of New York.

Ignoffo, M. J.
1991   Sunnyvale: From the "City of Destiny" to the "Heart of Silicon Valley." M.A. thesis. San Jose State University.

Institutional Planning and Academic Resources
2000   Statistical Abstracts. San Jose State University. Electronic document. http://www.ipar.sjsu.edu/factbook/statistical/99section7.html.

Janah, Monua
  1999 Journey to Success Began in Flight from Africa. San Jose Mercury News, July 18: 1E, 7E.
Johnson, Hans
  2000 Movin' Out: Domestic Migration to and from California in the 1990s. Public Policy Institute of California. California Counts: Population Trends and Profiles 2(1): 1–16.
Johnson, Hans, and Sonya Tafoya
  2000 Trends in Family and Household Poverty. Public Policy Institute of California. California Counts: Population Trends and Profiles 1(3): 1–12.
Johnson, Steve
  1992 Minorities Are the Majority in 137 Occupations in County. San Jose Mercury News, November 22: 1A, 19A, 20A.
Joint Venture: Silicon Valley Network
  1993 Blueprint of a 21st Century Community: The Phase II Report. June 1993. Sunnyvale: Consolidated Publications.
  1998 Silicon Valley 2010. San Jose: Joint Venture: Silicon Valley Network.
  1999 Workforce Study: An Analysis of the Workforce Gap in Silicon Valley. San Jose: Joint Venture: Silicon Valley Network.
  2000 The Joint Venture Way: Lessons for Regional Rejuvenation. Vol. 2. San Jose: Joint Venture: Silicon Valley Network.
  2001 Joint Venture's 2001 Index of Silicon Valley. San Jose: Joint Venture: Silicon Valley Network.
Jones, Sian
  1997 The Archaeology of Ethnicity: Constructing Identity in the Past and Present. New York: Routledge.
Jung, Carolyn
  1996a Local Execs Boost Manila "Silicon Valley." San Jose Mercury News, June 14: 1C, 3C.
  1996b Meet High Tech's Johnny Appleseed. San Jose Mercury News, July 1: 1E, 6E.
Kao, Henry S. R., and Sek-Hong Ng
  1992 Organizational Commitment and Culture. *In* Organisational Behavior: Southeast Asian Perspectives. R. I. Westwood, ed. Pp. 173–98. Hong Kong: Longman.
Kunda, G.
  1992 Engineering Culture: Control and Commitment in a High Tech Corporation. Philadelphia: Temple University Press.
Kuper, Adam
  1999 Culture: The Anthropologists' Account. Cambridge, MA: Harvard University Press.
Kvamme, E. Floyd
  2000 Life in Silicon Valley: A First-hand View of the Region's Growth. *In* The Silicon Valley Edge. Chong-Moon Lee, William Miller, Margue-

rite Gong Hancock, and Henry Rowen, eds. Pp. 59–80. Stanford: Stanford University Press.

LaFleur, Jennifer
1998   Education on Minds of Silicon Valley Voters. San Jose Mercury News, September 6: 1A, 19A.

Langberg, Mike, and Larry Slonaker
1997   House of Fry's. San Jose Mercury News, August 24: 1A, 12A–14A.

Legón, Jeordan
1995   The Name Game: Chicano, Hispanic or Latino—Labels Cause Division in Community. San Jose Mercury News, December 19: 1A, 12A.

Levine, Robert
1984   Properties of Culture: An Ethnographic View. *In* Culture Theory: Essays on Mind, Self, and Emotion. Richard Shweder and Robert Levine, eds. Pp. 67–87. New York: Cambridge University Press.

Lewis, Marilyn
1993   World Cultures Meet in Workplace. San Jose Mercury News, January 24: 1A, 22A.

Lewis, Marilyn, and Jeff Gottlieb
1993   New Demographics Fuel Ethnic Rivalry. San Jose Mercury News, May 31: 1A, 28A.

Lewis, Paul, and Max Neiman
2000   Residential Development and Growth Control Policies: Survey Results from Cities in Three California Regions. San Francisco: Public Policy Institute of California.

Lindholm, Charles
2001   Culture and Identity: The History, Theory and Practice of Psychological Anthropology. San Francisco: McGraw-Hill.

Liu, Philip
1991   Coming Home. Free China Review (December): 40–45.

Lyon, David
1998   The Wellsprings of California's Economic Growth: Myths and Realities. Occasional Paper. San Francisco: Public Policy Institute of California.

Mach, Zdzislaw
1993   Symbols, Conflict and Identity: Essays in Political Anthropology. Albany: State University of New York Press.

Malson, John, and Melanie Martindale
2000   City/County Population Estimates. California State Department of Finance. Electronic document. http://www.dof.ca.gov/html/Demograp/E-1text.htm.

Mangaliman, Jessie
2000   Indo-Americans Send Kids Back to Homeland to Learn Culture. San Jose Mercury News, October 30: 1A, 8A.

Markussen, Randi
1995   Constructing Easiness—Historical Perspectives on Work, Comput-

erization and Women. *In* The Cultures of Computing. Susan Leigh Star, ed. Pp. 158–80. Cambridge, MA: Blackwell.

Martinez, Anne
  1999   Assembly Language. San Jose Mercury News, August 11: 1B, 4B.
Matthews, Glenna
  1976   The Community Study: Ethnicity and Success in San José. Journal of Interdisciplinary History 7(2): 305–18.
McBrierty, Vincent, and Raymond Kinsella
  1998   Ireland and the Knowledge Economy: The New Techno-Academic Paradigm. Dublin: Oak Tree Press.
McCormack, Dan
  2000   McCormack's Guides. Santa Clara County 2000. Martinez, CA: McCormack's Guides.
McLaughlin, Ken
  1996   Torn between Worlds. San Jose Mercury News, April 14: 1A, 21A.
McLaughlin, Ken, and Ariana Cha
  1999   A Majority of None. San Jose Mercury News, April 14: 1A, 21A.
McLeod, B.
  1986   The Social Psychological Adaptation of Immigrant Chinese Professionals in California's Silicon Valley. Unpublished dissertation. University of California, Santa Cruz. Ann Arbor: University Microfilms International.
Miller, Bruce
  1988   Chumash: A Picture of Their World. Los Osos, CA: Sand River Press.
Mitcham, Carl
  1994   Thinking through Technology: The Path between Engineering and Philosophy. Chicago: University of Chicago Press.
Morgan, James
  1997   Valley's Hopes Rise with Jiang. San Jose Mercury News, March 2: 6C.
Muhammad, Tariq
  1996   Home Is Where the Hardware Is. Black Enterprise 26(8): 102–07.
Nash, June
  1989   From Tank Town to High Tech: The Clash of Community and Industrial Cycles. Albany: State University of New York Press.
Naylor, Larry
  1998   American Culture: Myth and Reality of a Culture of Diversity. Westport, CT: Bergin and Garvey.
Nhu, T. T.
  1997   Immigrants Have Driving Ambition. San Jose Mercury News, August 5: 1B, 2B.
Nippert-Eng, Christina
  1996   Home and Work: Negotiating Boundaries through Everyday Life. Chicago: University of Chicago Press.

Nolan, Riall
    1999   Communicating and Adapting across Cultures: Living and Working in the Global Village. Westport, CT: Bergin and Garvey.

Packard, David
    1995   The H-P Way: How Bill Hewlett and I Built Our Company. Edited by David Kirby and Karen Lewis. New York: HarperBusiness.

Park, Edward Jang-Woo
    1994   Asian Americans in Silicon Valley: Race and Ethnicity in the Postindustrial Economy. Unpublished dissertation. University of California, Berkeley.
    1996   Asians Matter: Asian American Entrepreneurs in the Silicon Valley. *In* Reframing the Immigration Debate. Bill Ong Hing and Ronald Lee, eds. Pp. 155–77. Los Angeles: Leadership Education for Asian Pacific (LEAP) and UCLA Asian American Studies Center. High Technology Industry.

Pendleton, Mary
    1975   Navajo and Hopi Weaving Techniques. New York: Macmillan.

Perkins, Broderick
    1998   Home Prices Highest in U.S. San Jose Mercury News, Feb. 12: 1A.

Perlow, Leslie
    1997   Finding Time: How Corporations, Individuals, and Families Can Benefit from New Work Practices. Ithaca, NY: ILR/Cornell University Press.

Pfister, Joel
    1997   Glamorizing the Psychological: The Politics and the Performances of Modern Psychological Identities. *In* Inventing the Psychological: Toward a Cultural History of Emotional Life in America. Joel Pfister and Nancy Schnog, eds. Pp.167–213. New Haven: Yale University Press.

Pickard, Naftoli
    1997   Negotiating Identities: A "Non-Techie" View from Silicon Valley. Senior honors thesis. San Jose State University, Department of Anthropology.

PRx, Inc., Strategic Marketing Communications
    1993   Joint Venture Silicon Valley Progress Report. San Jose: Pizazz Printing.

Public Policy Institute of California
    1999   Income Inequality in the Golden State: Why the Gap Has Widened between Rich and Poor. Research Brief. February 1999. Issue #17. San Francisco: Public Policy Institute of California.
    2000a California's Population. Electronic document. http://www.ppic.org/facts/cal.pop.dec99.pdf.
    2000b California's Digital Divide. Electronic document. http://www.ppic.org/facts/digital.nov00.pdf.

2000c Election 2000 Legislative Analysis. Electronic document. http://www.ppic.org/facts/legis.analysis.pdf.

Quinn, Michelle
1998 They Want the Recipe of Success. San Jose Mercury News, March 25: 1C, 2C.

Rae-Dupree, Janet
1995 Diversity Is Theme at Holiday Event. San Jose Mercury News, July 3: 1B, 4B.

Rodee, Marian
1987 Weaving of the Southwest. West Chester, PA: Schiffer.

Rogers, Everett, and Judith Larson
1984 Silicon Valley Fever: Growth of High-Technology Culture. New York: Basic Books.

Roseberry, William
1996 The Rise of Yuppie Coffees and the Reimagination of Class in the United States. American Anthropologist 98(4): 762–75.

Sandoval, Ricardo
1996 Training for Tomorrow. San Jose Mercury News, March 29: 1C.

San Jose Mercury News
1998a Jobless Rate Falls to 2.8% in Santa Clara County. April 14: 1C.
1998b 45 Years of Las Madres: Of Moms, for Moms, by Moms. July 9: 13E.

Saxenian, A.
1985 The Genesis of Silicon Valley. In Silicon Landscapes. Peter Hall and Ann Markusen, eds. Pp. 20–34. Boston: Allen and Unwin.
1994 Regional Advantage: Culture and Competition in Silicon Valley and Route 128. Cambridge, MA: Harvard University Press.
1999 Silicon Valley's New Immigrant Entrepreneurs. San Francisco: Public Policy Institute of California.

Sexton, Jean Deitz
1992 Silicon Valley Inventing the Future: A Contemporary Portrait. Hong Kong: Windsor Publications.

Sharp, Lauriston
1952 Steel Axes for Stone-Age Australians. Human Organization 11(2): 17–22.

Shweder, Richard, and Edmund Bourne
1984 Does the Concept of the Person Vary Cross-culturally? In Culture Theory: Essays on Mind. Self, and Emotion. Richard Shweder and Robert Levine, eds. Pp. 159–99. New York: Cambridge University Press.

Siegal, L.
1990 Silicon Valley's Workforce Remains Segregated. Global Electronics 101: 1–4.

Silicon Valley Manufacturing Group
1999 Population and Job Forecasts. Electronic document. http://www.abag.ca.gov/svprojections2000.pdf.

Smart Valley
    n.d.   Smart Valley Telecommuting Project. Electronic document. http://rpcp.mit.edu/diig/Related/smart_valley/tele_proj.txt.

SRI International, Center for Economic Competitiveness
    1992   Joint Venture: Silicon Valley, An Economy at Risk. San Jose: San Jose Metropolitan Chamber of Commerce.

Steen, Margaret
    2000   Jobs Still Abound. San Jose Mercury News, October 14: 1C, 4C.

Stocking, Ben
    1999a   Changing Face of the Future. San Jose Mercury News, April 18: 1A, 22A–24A.
    1999b   Cupertino Adjusts to Influence of Immigrants. San Jose Mercury News, April 15: 1A, 14A.

Sullivan, Richard
    1979   The Medieval Monk as Frontiersman. In The Frontier: Comparative Studies. Vol. 2. William Savage and Stephen Thompson, eds. Pp. 25–49. Norman: University of Oklahoma Press.

Tafoya, Sonya
    2000   Mixed Race and Ethnicity in California. Public Policy Institute of California. California Counts: Population Trends and Profiles 1(2): 1–12.

Taylor, Charles
    1985   Philosophy and the Human Sciences. Philosophical Papers 2. New York: Cambridge University Press.

Tech Museum of Innovation
    n.d.   Our Mission. Electronic document. http://www.thetech.org/ops/mission.html.

Teraguchi, Maho
    1997   Chuzaiin—Visiting/Sojourning Japanese Business People in Silicon Valley. Paper presented at the Biennial Meetings of ASPAC. Asilomar, CA, June 18.

Textor, Robert
    1985   Anticipatory Anthropology and the Telemicroelectronic Revolution: A Preliminary Report from Silicon Valley. Anthropology and Education Quarterly 16: 3–30.
    1995   The Ethnographic Futures Research Method: An Application to Thailand. Futures 27(4): 461–71.

Thoden van Velzen, H. U. E.
    1990   The Maroon Insurgency: Anthropological Reflections on the Civil War in Suriname. In Resistance and Rebellion in Suriname: Old and New. Studies in Third World Societies. Publication 43. Gary Brana-Shute, ed. Pp. 159–88. Williamsburg, VA: College of William and Mary.

Thorne, Avril, and Harrison Gough
    1991   Portraits of Type: An MBTI Research Compendium. Palo Alto: CPP Books.

Thurm, Scott
    1996   Valley Defied Growing Income Gap. San Jose Mercury News, August 4: 1A, 25A.

Thurm, Scott, and Patrick May
    1997   How Deep Does the Boom Go? San Jose Mercury News, July 20: 1A, 20A.

Tiles, Mary, and Hans Oberdiek
    1995   Living in a Technological Culture. New York: Routledge.

Tran, De
    1998a   World Gets Day on Stage. San Jose Mercury News, April 30: 5B.
    1998b   The New Village Green. San Jose Mercury News, April 5: 1B, 5B.

Tran, Tini
    1996   Classrooms Make Connection. San Jose Mercury News, October 13: 1B, 2B.

U.S. Census Bureau
    n.d.   Detailed Occupation by Race, Hispanic Origin and Sex: Janitors and Cleaners. Electronic document. http://tier2.census.gov/cgi-win/eeo/eeodata.exe.
    2001   State and County QuickFacts. Santa Clara County. Electronic document. http://quickfacts.census.gov/qfd/states/06/06085.html.

van den Berghe, P. L.
    1967   Race and Racism: A Comparative Perspective. New York: John Wiley.

van der Elst, Dirk
    1970   The Bush Negro Tribes of Surinam, South America: A Synthesis. Ph.D. dissertation. Northwestern University.

van der Elst, Dirk, with Paul Bohannan
    1999   Culture as Given, Culture as Choice. Prospect Heights, IL: Waveland Press.

Wasserman, Elizabeth
    1996   Where the Bucks Are. San Jose Mercury News, November 17: 1A, 21A.

Weimers, Leigh
    1998   Silicon Valley Charity Ball. San Jose Mercury News, April 19: 1B, 4B.

Wilkins, Teresa, and Diana Leonard
    1990   Beyond the Loom: Keys to Understanding Early Southwestern Weaving. Boulder: Johnson Books.

Winner, Langdon
    1992   Silicon Valley Mystery House. *In* Variations on a Theme Park: The New American City and the End of Public Space. Michael Sorkin, ed. Pp. 31–60. New York: Noonday Press.

Winslow, W.
    1995   The Making of Silicon Valley: A One Hundred Year Renaissance. Palo Alto: Santa Clara Valley Historical Association.

Youssouf, I. A., A. D. Grimshaw, et al.
   1976   Greetings in the Desert. American Ethnologist 3(4): 797–824.
Zloniski, Christian
   1994   The Informal Economy in an Advanced Industrialized Society: Mexican Immigrant Labor in Silicon Valley. Yale Law Journal 103: 2305–35.

Index